The Aesthetics of Industrial Design

This textbook introduces design students to key principles of three-dimensional form, bridging aesthetics and practical design objectives. It explores how we see and what it is that characterises visually appealing and satisfactory design.

Written by an experienced designer, educator and researcher, *The Aesthetics of Industrial Design* equips students with the knowledge and understanding of how aesthetically superior design is distinct from lesser work. It explains the key principles and concepts they can incorporate into their own designs, encourages readers to investigate and experiment with real design problems and enables them to verbally communicate their design intentions. The book prompts readers to critically reflect on their work and surroundings. Through numerous clear examples and illustrated case studies, which are guided by cognitive science and the application of aesthetic theory, the book brings together the basic aspects of design as form-giving. It explores the balance of function, material and appearance in detail and explains the reasons for common aesthetic faults and how to avoid them.

Aimed at undergraduate- and postgraduate-level students within the design fields, this book reveals the secrets to aesthetically successful products that readers can take from education into future practice.

Richard Herriott is Associate Professor in Design for People and Industrial Design at Design School Kolding in Denmark.

"This book covers a wide range of issues which are important for the future of design. The fact that the author gives tasks throughout the book is a helpful feature. I appreciate that the author encourages discussions when things don't easily fall into categories or are more complex so there are no simple answers. I believe this book, which attempts to merge theory and practice, is important to publish. It gives us all a possibility to discuss the complexity of such an undertaking and start discussions about how we formulate basic aesthetic principles that are free from judgement and when value judgements are made that apply principles in specific design processes."

Cheryl Akner-Koler, Professor, Department of Design,
Interior Architecture and Visual Communication/Industrial Design,
University of Arts, Crafts and Design, Stockholm, Sweden

"A much-needed and highly lucid exposition of the thought and technique behind successful industrial design for students of this and allied professions. Endearingly opinionated and idiosyncratic in places, the book offers a path from professional philosophy to thoughtful execution of high-class industrial design work. Useful exercises and debating points are provided for the benefit of tutors and students alike."

Peter Barker, Head of School, Design and Communication,
Plymouth College of Art, UK

"This book is a brilliant example of why Kurt Lewin was right in stating that there is 'nothing as practical as a good theory' (1943). Through a row of examples of good, functional, robust, aesthetically appealing and beautiful design as well as 'design mistakes' and 'bad designs', the author, associate professor Richard Herriott from Design School Kolding takes us through theories about visual perception, constraints, lines, surfaces and curvature, product semantics and craftsmanship and discusses the implications of these theories for design. Even if some of the theories are complicated reading, Herriott takes the reader by the hand and illustrates the theoretical concepts and terms via reference to examples from real-world design, models and drawings. The book covers an impressive number of ideas and invites the reader to think about and re-think design as well as reflecting upon the role of the designer in the contemporary design practice. Doing this, the book is a great vehicle for designers aiming to create and analyse objects that are aesthetically pleasing and for others to understand the science of forms and the building blocks of good design."

Lene Tangaard, Rector of the Kolding Design School, Denmark

"Disagreeing with Richard Herriott is a source of great pleasure. Not because standing at the opposing end during an argument lends one a feeling of superiority – far from it – but because few people can make a point the way Richard does. His immense knowledge provides him with references that are, at the very least, curiously entertaining and at best enlightening. His intellect then allows Richard to process these select references in a highly original a manner. I'm delighted that this book provides many people with the opportunity to sneak a peek into Richard Herriott's singular mind. They may agree or disagree with what they find – but they will certainly be all the wiser for it."

Christopher Butt, writer and design critic,
founder of Auto-Didakt and Design Field Trip

"The human brain is a mystery. Barely developed since mankind evolved itself into this peculiar world, we still react on basic instincts and pre-coded assumptions. We think we know, but we don't. We think we're in control, but we're not. So we need to learn. And of all our senses, our visual perception is our least reliable. Richard Herriott has created a fantastic book that enlightens us and opens our mind to understand better how our brain acts and reacts to our visual surroundings, such as form, function, design and visual perception in general. In a well-written and richly illustrated piece, he guides us through various elements in modern design such as the human visual system, the complexity in form and function, how we look at and understand shapes and form, etc. Starting with the already known theories, Richard Herriott describes and discusses the complexity and compromises we make daily in our visual world. Wisdom and knowledge mixed with simple illustrations and easy-to-use exercises. It's a must-read for everybody working with visual art forms, teachers, designers, photographers, architects, artists etc., a need to read for everybody interested in how our brain works when it comes to visual perception – and a nice to read for everybody else that just want to know. A welcoming and necessary book in a time where everything runs so fast that we tend to forget what we already know."

Kåre Birk, teacher in Furniture, Space and Products,
Scandinavian Design College, Denmark

The Aesthetics of Industrial Design

Seeing, Designing and Making

Richard Herriott

Routledge
Taylor & Francis Group

NEW YORK AND LONDON

Cover image: © Joni Mok

First published 2022
by Routledge
4 Park Square, Milton Park, Abingdon, Oxon OX14 4RN

and by Routledge
605 Third Avenue, New York, NY 10158

Routledge is an imprint of the Taylor & Francis Group, an informa business

British Library Cataloguing-in-Publication Data
A catalogue record for this book is available from the British Library

Library of Congress Cataloging-in-Publication Data
Names: Herriott, Richard, author.
Title: The aesthetics of industrial design : seeing, designing and making /
Richard Herriott.
Description: Abingdon, Oxon ; New York. NY : Routledge, 2022. |
Includes bibliographical references and index.
Identifiers: LCCN 2021040338 (print) | LCCN 2021040339 (ebook)
Subjects: LCSH: Industrial design--Textbooks. | Aesthetics,
Modern--Textbooks.
Classification: LCC TS171.4 .H47 2022 (print) | LCC TS171.4 (ebook) |
DDC 745.2--dc23/eng/20211008
LC record available at https://lccn.loc.gov/2021040338
LC ebook record available at https://lccn.loc.gov/2021040339

ISBN: 978-1-032-02419-6 (hbk)
ISBN: 978-1-032-02418-9 (pbk)
ISBN: 978-1-003-18330-3 (ebk)

DOI: 10.4324/9781003183303

Typeset in Univers
by Deanta Global Publishing Services Chennai India

For my parents, with all my love, for all their support down the years.

Contents

List of Illustrations   x
Acknowledgements   xxiii

1   How do we see?   1

2   Resolving the constraints   50

3   Lines, surfaces and curvature   76

4   Craftsmanship   120

5   The meaning of the object and its elements:
     Product semantics   162

References   193
Index   197

Illustrations

FIGURES

1.1   This is not a tram. Titled with due deference to René Magritte   2

1.2   Schematic model of part of macaque monkey's visual system.
The detail is not as important as the apparent complexity of
the system (from Grossberg, 2014)   3

1.3   Overview of the themes of Chapter 1   5

1.4   Apartments and bus stop, Aarhus, Denmark. Individually the
elements are at least tidy and neat, but as a whole, it is badly
designed (or not designed at all). Bad design seems to involve
a lack of control or loss of control regarding how the totality
of the ensemble is perceived. I had originally picked an ugly
Belgian house for this illustration but decided that it was an
artistic statement and not bad design at all. The search for
ugly or badly designed things led me to realise that ugliness
is often about context: things that don't fit in or don't work
together. The sheer ghastliness of many urban settings (design
that is bad) is often not easily photographed. But being there
is demoralising because the place shows no concern for the
feelings of residents. It is mere engineering. Architects and
planners have been much more prolific than designers in
making sad and bad things. At least with consumer goods you
can avoid them or dispose of them if they are too nasty   6

1.5   Figure-ground. On the left, the edge belongs to the black
shape, which is a black figure on a white background. The
middle image is a figure on a ground. What is going on with
the image on the right, though? The kettle is a figure on
a ground *and* the photograph of the kettle is a figure on a
ground (the page). Which do you see most strongly?   9

1.6   Figure-ground. In this instance a white oblong can seen as a
hole in a black triangle. The edge now belongs to the black
triangle, which is sitting on a white background. You can also
actively choose to see the white area as a box on a black
background   9

1.7   Proximity: these are seen as two groups, not 24 individuals   10
1.8   The area 'A', outlined with a box, is seen as one group. The
       six other buttons are perceived as two subsidiary groups. The
       arrangement makes sense given their functions   10
1.9   Similarity: the grey squares are seen as units (rows)   11
1.10   Common destiny. We can see the bubbles arising from the
       pipe as grouped as they look as if they are heading the same
       way. The single bubble is not seen as part of the group of
       bubbles. It looks as if it is heading in another direction. The
       noticeability of the one bubble is marked   11
1.11   Law of Simplicity: these 17 curves are seen as five ovoids   12
1.12   This telephone is seen as one item even though it is not   12
1.13   Factor of direction. The points A and B and points C and D are
       seen as being connected by a curve each. Less likely is that
       there are four curves A to F, F to B, C to F and F to D   13
1.14   The factor of closure. One interpretation is that this diagram
       (left) shows three separate lines. Another is that it is a single
       square, partially obscured. In the right image we sense that
       the camera is in the way of the box under it   13
1.15   Factor of closure. In this case it is not much of a jump to see
       this as a square. It is hard not to   13
1.16   Factor of closure. In this case the impression of a square is
       very weak. The gaps are large relative to the remaining lines
       and the corners are missing   14
1.17   Factor of closure. In this case, although the gaps are large
       relative to the lines, the presence of the corners strongly
       reinforces the impression of a square. There is also the weak
       impression that a cross-shaped form is interrupting the view
       of the square   14
1.18   Factor of good curve? This shows an ambiguous design with
       several laws in conflict   15
1.19   The same car seen from the side. The line AB is suggested by
       the internal structure of the lamp. The line BC is a compound
       of the glass edge and a black line in the lamp. The shape D is
       the outline of the lamp. So, is the lamp a figure on a ground?
       It is ambiguous   15
1.20   How gestalt theory and the theories of Gibson and Gregory
       are related   17
1.21   In the left diagram the three circles are interpreted as having
       different sizes. The lowest one is seen as smallest. The top left
       one is the largest and is understood as furthest away. In the
       middle diagram, the compression of lines is seen as showing
       increasing distance. The right diagram is interpreted as showing
       parallel lines receding. The lines are not actually parallel   18

1.22   A box of socks   22
1.23   There is a lot to take in here. It's a ticket machine with a lot of
        surplus information (in Gregory's terms) and the affordances
        aren't clear (Gibson's terms)   23
1.24   This image show a similar effect to the spinning hollow
        mask phenomenon. It shows the impression of a face in
        snow. It can be seen as a hollow form (concave) or as a form
        projecting out from the snow (convex). It depends on what
        you want to see: what you know is there and what you think
        is there. (Image: Wiki commons)   24
1.25   Necker cube. Which square is nearer?   25
1.26   Penrose triangle   26
1.27   A simple drawing suggesting distance and space   27
1.28   KM3 Food Processor (1957) Mixer by Gerd A. Müller for
        Braun. (Image courtesy of MAKK – Museum für Angewandte
        Kunst Köln, Germany.)   28
1.29   This isn't a face   30
1.30   An example of a car design exploiting anthropomorphic
        effects. This car can be viewed as being cute because of its
        large, round lamps and smiling grille   31
1.31   Ox chair (left, designed by Hans J. Wegner and manufacturer
        Erik Jørgensen, Fredericia Furniture, Denmark) and Swan
        (right, designed by Arne Jacobsen). (Image: Wiki Commons)   33
1.32   Design action in relation to an object's perceived gender   34
1.33   This shows a house sketched from a primitive form. Our
        visual understanding of the house may be the reverse
        whereby we first perceive an oblong and then the details   36
1.34   A small child's representation of a car   37
1.35   Many artificial objects can be drawn using these forms
        ('primitives') as guiding frameworks   38
1.36   Picture 1 is a volume such as a box. Picture 2 shows an axis
        A–B that summarises the box's character. Picture 3 shows
        an axis with movement (deformation) compared to the box in
        picture 1. Picture 4 visualises the movement (deformation) as
        the result of a force C. We see the box in picture 4 as if it has
        been deformed   40
1.37   Box A can be seen as a version of box B   40
1.38   Box A seems to have been cut, compared to 'ideal' box B   40
1.39   The vertical lines on the left side of the box seem to be
        continuous. It looks as if something was cut away   41
1.40   A block (left) subject to mild deformation (right)   42

1.41 A simple oblong with its primary (1), secondary (2) and tertiary (3) axes. Is there a gestalt law in operation in this image? Yes. The Law of Continuity – we see the axes as being continuous despite the box in the way. We infer the primary axis is one line, passing through a box 42

1.42 Deformed plane. Redrawn from Akner-Koler (1994) 42

1.43 Another interpretation of the deformation: an even force is applied to the left edge, moving it up relative to the right edge 43

1.44 This Pontiac Trans Am is strongly directional 43

1.45 Bent (A) and curved axial deformation (B). A is a bending force and B is curving force 44

1.46 The elements of the chair can be described in terms of forces 45

1.47 Forces' direction and effect. Redrawn from Akner-Koler (1994) 46

1.48 A shape with an axis which is curved. The shape of the axis is an expression of the outward shape's overall form 46

1.49 An electric kettle with its subtle accents expressed in terms of axial forces. Which notion from Gibson is in operation here? It's the notion of affordance. The forces are supporting the idea of the kettle's motion upward and of holding the kettle 47

2.1 How the elements of this chapter are related 51

2.2 There are two fairly decent curves here (and one not so good). Deciding which one to use means more than only being able to assess curvature quality in isolation 52

2.3 Everything affects everything else: cost, quality, appearance, functionality. Material is not explicitly included. It is here subsumed under appearance (for the right look) and quality (for the right durability). What means exists to assign priority to these demands? A design process is how one finds a path through the conflicting demands of a project 53

2.4 The design of this wooden-cased 1960s radio is strongly related to cutting processes. (Image: author's collection) 55

2.5 This Apple plug is made from injection moulded plastic. Its aesthetic is related to a cheaper and simpler production process, but note the high quality of the curvature of the corners and edges 55

2.6 Tour Vercors, Grenoble (1966). In this case architects have decided to show the living modules as distinct units. The wish to for multi-story construction determines the material. It could not have been made of brick, wood or stone and still looked like this 56

2.7   The remote control (left) makes no obvious concessions to
        ergonomics. The visual order is the sole aid to the user. The
        telephone (centre) has quite large buttons with clear labelling
        so the basic functions are clear. The device is also slightly
        narrower at the end to afford holding it (it is narrower at the
        top end to look symmetrical). The mouse (right) is subtly
        ergonomic – it is smooth to fit in the palm and has a thumb rest   60
2.8   The close relation of the building footprint to the carriageways
        means this street has visible affordances as to the directions
        pedestrians and drivers can go. Some of them are marked up
        with arrows. Are there any affordances that could be added?   60
2.9   Wassily chair by Marcel Breuer. (Courtesy of Knoll, Inc.)   61
2.10   Decorative chair, c. 1920   62
2.11   A contemporary hairdryer. (Courtesy of Braun)   64
2.12   A plastic part coated to look like alloy   66
2.13   A material and form decision-making flow chart. For it to
        work you need to know what shapes are consistent with the
        material in question. You could also ask users at the very start
        whether they are happy with the material and the form. Many
        users probably won't mind very much. This kind of issue is of
        more concern to professionals and critics!   68
2.14   Georgian handmade tableware (left) and a mass-produced
        vacuum cleaner (right). (Left image: Credit: By Sean
        Pathasema/Birmingham Museum of Art, CC BY 3.0, https://
        commons.wikimedia.org/w/index.php?curid=18805475)   70
2.15   Victorian interior. (Image: Wiki Commons)   72
2.16   Function and form spectrum   74
3.1   Mind map of the contents of this chapter   77
3.2   There are mostly smooth transitions between sections of the
        curve A to H. But at D there is an abrupt change from one
        curve section to the next. Curve 1 shows a marked bump.
        Curve 2 shows a flat area in the middle. It looks ambiguous   78
3.3   *Snake* by Phil Price, Aarhus, Denmark. Note the smooth
        transitions between sections of the work. The aesthetic effect
        of the structure depends almost very much on the gentle
        curves being smoothly blended   79
3.4   Staffordshire Tableware mug (left); Dibbern mug (batch 6205)
        (right). (Author's collection)   81
3.5   The feedback from the concept to the drawing to the model.
        Puzzles at the drawing stage can be related back to the
        concept, which can be revised, and puzzles in the model
        stage can be worked over using new drawings and revising
        the concept until all three are in equilibrium   84
3.6   Conceive, critique and control   86

3.7   Axial space. Z is up, X is length and Y is width   88
3.8   The straight line in space. It is situated on the plane of Y
       where Y equals zero and has a height difference on the Z axis
       (A is higher than B). B is further away from the origin 0,0,0
       where X and Y and Z are all 0   88
3.9   A curved line connecting points A, C and B   89
3.10   The arc in space. It is still two-dimensional. The 'snake' in
       Figure 3.3 is strongly three-dimensional   89
3.11   The arc in space. It is still two-dimensional. The curve shows
       more activity near the 'C' than closer to the 'B', where it is
       flatter or less active   90
3.12   The curve has now been extended into three-dimensional
       space. It is now a surface with four edges. There is a gestalt
       principle operating in this image now. What is it? Continuity.
       The Z axis and X axis lines are interrupted by the surface, but
       one assumes they are still there, hidden   90
3.13   An uneven surface and a section, A–B, showing the wobbles
       in the surface (left). An even surface with smooth curvature
       (right). A section A–B showing a clean, smooth line (bottom)   92
3.14   A simple arc   92
3.15   A circle is a closed line with constant curvature   93
3.16   This looks like a flattened circle. It has more curvature at
       the sides than at the top. The rate of curvature increases or
       accelerates from the top part to the sides. From the sides
       down, the curvature decreases again   93
3.17   The arrows show the direction of accelerating curvature
       towards the middle of the surface. The surface has an axis of
       curvature parallel to the Y axis   94
3.18   Gradual acceleration in curvature from the centre to the corner
       marked A–B. The curvature is most pronounced between A
       and B and then diminishes as the line goes downwards again   94
3.19   Highlights from overhead strip lighting on a car bonnet   95
3.20   *Waiting Swallow* (2017), Ian Pollock   95
3.21   *Beak* jug by Georg Jensen, Denmark   96
3.22   *Beak* jug by Georg Jensen, detail of handle and body junction   96
3.23   A rounded edge. Probably a third of the time spent modelling
       a complex form might be taken up doing rounded edges. It
       can get tricky where the fillets meet up   98
3.24   A generalised diagram of two main surfaces and a transitional one   98
3.25   A simple rectangular monovolume (top left). A compound
       form with a primary and secondary volume (bottom left). On
       the right side are versions with rounded edges added   99

3.26 Main surfaces A and B, secondary surfaces C and D and their associated fillets 1–6. Notice that the fillets 1–6 are smaller than the radius of the tertiary surface between C and D. The fillets 1–6 dominate the secondary surfaces C and D and the tertiary surface between them 100

3.27 A pair of curves with a gap (left) and two ways of blending them below (top left). Either is possible but both should be smoothly blended to the main curves. The same curves in 3D (right). Joining the two sides is about finding a smooth blending surface 100

3.28 The phone again. Two monovolumes from Chapter 1, the body of the phone and the handset 101

3.29 Fillet combinations on a rectangular form 102

3.30 Smaller fillets but the same underlying geometry as in Figure 3.29 102

3.31 Combinations of fillet sizes and curves/sections of surfaces. A section through two flat main surfaces with a small fillet (top left). Two main flat surfaces and a very dominant, large fillet (top right). How large fillets on very curved surfaces produces a very inflated-looking shape (bottom right) 103

3.32 A positive space (a volume) on the left; a negative space (inside the bowl) on the right 105

3.33 A form constructed of three main surfaces. The light source is located above the object. Notice how the shadow falls on each panel 105

3.34 Drawing (2018) by Lucian Bové, Renault Advanced Design Centre 106

3.35 Side profile of clay model of 2011 Ford Focus. (Image courtesy of Ford, Europe) 106

3.36 Side profile of clay model of 2011 Ford Focus, with shadows marked out 107

3.37 The lower oblong is perceived as being longer than the upper one, more so when seen in isolation. In fashion, stripes on shirts are usually vertical, to make the wearer look a bit thinner 107

3.38 A rough sketch of how a cylindrical form with rounded ends reflects a grey area below it 108

3.39 A rough sketch of how a cylindrical form with rounded ends reflects a grey area below it and a lighter one above it. Look closely at an object near you and see where the light and shadow fall 108

3.40 Clay modelling exploring the relation of light and shade. What is the effect of those bands of light and dark? (Image courtesy of Ford, Europe) 109

3.41 Digital radio with some rounded corners 111

3.42 2011 VW Up! – Apparently simple forms which are really very refined indeed 112

3.43 Sketch section of chamfer feature on the Up! car (Figure 3.42). It's the way the curvature is handled on the chamfer that makes the reflections behave as they do. The hollow/convex/negative form of the surface and its upward-facing angle ensure it catches the light under most conditions 113

3.44 Electrolux UltraSilencer Pia Wallen edition, with joint lines (left) and without joint lines (right). Are the dots form or are they graphics? (Image: Electrolux) 114

3.45 Sketch of a plane intersecting a curved surface 115

3.46 Two irregular surfaces intersecting. The curve resulting from this is also uneven and will have changes in curvature that look disturbing. Is it supposed to be a line or a simple curve? It is ambiguous. You would not draw a curve like that. This kind of thing crops up when you try to turn a sketch into a 3D model and the intersection of two forms do not (in CAD) yield a good intersecting curve. Another, more typical hazard is where two surfaces intersect to make a curve that is a bit flat somewhere along its length 115

3.47 Intersection of one or possibly two uneven surfaces 116

3.48 Applied graphics on a vacuum cleaner. The coloured elements can also be considered as graphic elements. Notice also the way the part joints correspond to the coloured areas. Colour areas that cross part joint lines need to be painted coatings or applied films and are costly to do and prone to abrasion during use 118

4.1 How the elements of the chapter relate to each other. Semantics is dealt with in the next chapter 121

4.2 The landscape of joints and junctions on an electric kettle, as seen from 5-cm distance 122

4.3 The tectonic triangle, after Frampton (1995) 123

4.4 A building seen as a whole (left) and a product seen as a whole (right). Arguably, the nature of the joints in the building is not as critical to the overall perception of the form as it is in a product. (Building image: Wiki Commons. Product design sketch courtesy of Kitchen Innovations and Spark Innovations, Canada) 124

4.5 Detail of Lighthouse apartment complex, Aarhus, Denmark. Many small-scale components dominate at human scale, particularly the joints. Is this what was shown in the planning application and advertising material? 125

4.6 Exploded view of object by Kim Nicolaysen, Green Design, Denmark 128

4.7　　A sketch of a three-part item (left) and two alternative
　　　　principal sections (middle and right) that illustrate how the
　　　　item might be assembled. Making sketches like this requires
　　　　you to confront the details that can help or hinder the way
　　　　the user perceives the object. Notice particularly how part 3
　　　　relates to part 2　　　　　　　　　　　　　　　　　　　　129

4.8　　Pantone chair (1965). (Image: Wiki Commons)　　　　130

4.9　　The front bumper of an inexpensive car. Note the grooves
　　　　under the headlamps which suggest separate parts even
　　　　though it is a one-piece moulding　　　　　　　　　　131

4.10　A sketch of 'clamshell' type of box (top). Notice the section
　　　　a-b (middle). At the bottom is a section through the meeting
　　　　of the upper and lower parts. It shows failure types 2, 3 and
　　　　4. In number 1 the joint is successful: a flush fit. In the others,
　　　　the parts don't meet successfully, and 2 and 3 are offset
　　　　horizontally. In 4 the top doesn't land on the bottom part so
　　　　there is a gap　　　　　　　　　　　　　　　　　　　　132

4.11　Basic joint solution　　　　　　　　　　　　　　　　133

4.12　An approximate sketch of the joint concept in Figure 4.11 with
　　　　small fillets and stepped landing surface　　　　　　　134

4.13　Butt joint (left). The alignment of A and B might not be
　　　　adequate; in the diagram (right) the vertical is not aligned to
　　　　the horizontal. What would a solution to this be? One answer
　　　　is to have a raised lip around the area where the vertical part
　　　　meets the horizontal. That would make variations in fit harder
　　　　to see. On the right, the mitre joint, which requires more
　　　　precision. The extra difficulty adds to its appeal　　　　135

4.14　Train window pillar with butt joint; general context　　135

4.15　Train window pillar. Notice the uneven gap　　　　　　136

4.16　Butt joint (left) on the corner of a storage unit. It is not flush
　　　　and it is not aligned. The curvature is not that consistent
　　　　either. A mitre joint on the window frame of a 1982 Volvo 760
　　　　(right) (designed by Jans Wilsgaard)　　　　　　　　136

4.17　Sleeve joint. The distance A to B can vary according to the
　　　　location of the mounting points　　　　　　　　　　　137

4.18　Shingle joint. The overlap A to B can accommodate
　　　　variation in the parts' dimensions and also the location of
　　　　the fixing points　　　　　　　　　　　　　　　　　　137

4.19　Shingle joint on a window ledge　　　　　　　　　　　138

4.20　Photocopier exterior panels. Notice the variation in the panel
　　　　gaps and shut lines　　　　　　　　　　　　　　　　　139

4.21 A handle on a photocopier panel. The gap parallel to the hinge (on the left) is wider than that at the ends. Is there anything else that might be troublesome in the way this handle and its recess has been designed? The rounded corners on the left are sharper than those on the right. Maybe they should have been the same on all four corners. The matter is discussed further in Section 4.3   139

4.22 Collar joint on a food blender. It's the ring-shaped area between the brushed steel panel and the black plastic rotary dial   140

4.23 Intersection of a closed form (here it is a cylinder) and flat surface is in principle the essence of a collar joint   141

4.24 A simple collar joint. This intersection can be articulated in many ways   141

4.25 A schematic layout for panel gaps   142

4.26 The lines with fillet and gap applied. Notice the small triangular areas at each junction (1, 2 and 3)   143

4.27 Where the lines D meet the oval lines there is a quite large triangular gap which is unsightly. The curvature at F also needs attention   143

4.28 This is how the schematic lines in Figure 4.27 would look as a 3D model. Notice in the right view that there is a conspicuous gap where the linear gap meets the oval gap. (Model: Patrick Bomholt)   144

4.29 The circled area shows a three-way junction between a car headlamp, the bonnet and wing. Do you think the elimination of the radius on the upper side of the junction was a good solution?   144

4.30 The lines D meet the oval shape after swerving to intersect at something like right angles (g'–h')   145

4.31 This is how it would look at as a 3D model. Notice how the triangular junction is now less conspicuous. (Model: Patrick Bomholt)   145

4.32 This example is analogous to the schematic drawing in Figure 4.30. The panel gap between the bumper at the front wing intersects the wheel arch at A. This kind of thing is not easily shown in sketches and is often resolved at the modelling stage. The designer needs to be aware of this kind of problem and to anticipate such matters as early as possible during ideation   146

4.33 The gap c–d is the minimum permitted distance between the left side and right side of the gap; due to the edge having to be open for drafting purposes, the gap at the top is wider. The fillets increase this effect by making edges harder to perceive   146

4.34   Lines intersecting with a circle. Notice the apparent thickening
       of the circle at the points of intersection ('blips' or 'splotches'
       or 'blobs')   147
4.35   How the circular part of the form in Figure 4.34 could be
       interpreted. Following gestalt theory, the outline belongs to
       the circle. The intersecting lines seem make the outline look
       uneven   148
4.36   Detail of interior of a Deutsche Bahn ICE train. The vertical
       part is about 12 cm across. Notice the flaring of the trim on
       the upper right and also the visually constant gap condition   149
4.37   Interior detail of Danish regional train. Notice the visible screw
       heads and sleeve joints   150
4.38   Parallelism. The upper box is slightly off-parallel. The right
       side and left side are not the same length. This condition is
       ambiguous and is rectified by adjusting the length so the left
       and right sides are equal   151
4.39   Controlled convergence. The upper box is slightly off-parallel.
       The right side and left side are not the same length. This
       condition is ambiguous and is rectified by adjusting the
       lengths so that it is clear that one side is shorter and the
       asymmetry can be read as an active choice   151
4.40   Unaligned rectangles. The objects are close to alignment and
       identifiable as a group. However, the difference is not marked
       or obviously intentional. This could be a set of buttons or
       some windows   152
4.41   One solution is to bring all of the rectangles to a smaller
       common size, between the lines A–B and C–D   152
4.42   The alternative is to make all of the rectangles conform to the
       largest dimension between lines A–B and C–D   152
4.43   The connecting part between the two vertical elements is
       mathematically adequate but looks frail in relation to the pillars   154
4.44   A stronger-looking solution. There is no doubt this will carry
       people from one side to the other   154
4.45   A nice, neat oblong with rounded corners. Modelling this
       with nice radii on the corners and making sure it's all aligned
       and tidy could take a while if it was set on a double curvature
       primary surface   155
4.46   Seen in context, the larger form is not consistent   155
4.47   Door opening lever on a car door. The lever and its bezel are
       neat bits of work   156

4.48   Door opening bezel and its surroundings, marked up. If
        you look at the image in a general, nonfocused way, the
        handle and bezel seem to float unrelated to the surrounding
        elements. The bezel seems to have been stuck on rather than
        being an integrated element of the door panel   157
4.49   The path forward from two unsettling design options. Key
        here is that when you are confronted by two or more bad
        choices, don't be trapped into choosing one before examining
        the constraints more closely   160
5.1   Why product semantics?   163
5.2   How the elements of this chapter are related. The central
        part seems to be the idea of the product as a channel for
        communication. Successful products have a clear message
        for the user   164
5.3   Two raincoats. They differ in their use of product semantics.
        Which one is for traditionalists? (Left: courtesy of Barbour;
        right courtesy of WantDo Inc.)   165
5.4   The semantic triangle (adapted from Krippendorff and
        Butter, 1984)   166
5.5   The product as a sign   167
5.6   'No smoking' icon   168
5.7   Portable radios/hi-fis. Denver DAB 38 (left) and Lenco SCD
        420 (right). (Images courtesy of Denver and Lenco)   171
5.8   The semantic triangle and material applied to the radio   171
5.9   This diagram also shows where confusing signals can come
        in. (Adapted from Monö, 1997)   172
5.10   Shows the ways in which a product message gets lost along
        the way from the first idea to production. (Adapted from
        Monö, 1997)   173
5.11   Kettles from Japan (left) and the United States (right). (Left
        image courtesy of Naoto Fukasawa Design Ltd. Photo by
        Hidetoyo Sasaki)   173
5.12   In this diagram the elements of the 'channel' from Figure 5.9
        are expanded out. The main part of the diagram (adapted
        from Kicherer, 1987) shows all of the elements that add up to
        a perception of form. The 'gestalt' is the whole object (from
        the German word for 'a whole entity')   175
5.13   The relation of style to semantics in a revised version of
        Kicherer's conception of product semantics. Style is now an
        umbrella term for the elements connected to its right   176

5.14   Conceptual model of the Offenbach theory of product
        language (modified from Gros, 1976)   178
5.15   Conceptual model of the Offenbach theory of product
        language: the formal aesthetic function   179
5.16   Conceptual model of the Offenbach theory of product
        language: the indication functions   180
5.17   Conceptual model of the Offenbach theory of product
        language: the symbolic functions   181
5.18   The connection between the partial style concepts and
        the associations concepts. Notice that the diagram shows
        concepts as being more clearly separable than they really are.
        That is not a failure, just a reflection of reality's complexity.
        Designed objects' meanings are multi-layered and the layers
        are not always distinct   182
5.19   Gathering and using information relevant for a product's
        semantics   183
5.20   Contemporary building, Aarhus, Denmark, circa 2010   187
5.21   Grundig type 2029 radio, 1960   187

TABLES

1.1   Comparison of the rapid process of indirect perception with
        hypothetico-deductive reasoning   21
2.1   Some materials, their typical processes and typical
        resultant forms   65
3.1   The quantitative part is basic but might only be determined
        later, when the saltiness, volume or curvature come up for
        discussion. The intersubjective part is what most people
        would agree on; the subjective part is where judgement and
        taste come in   84

Acknowledgements

Thank you to Joni Mok for the process diagrams and cover photography, Patrick Bomholt for the CAD model images and Daniel O'Callaghan for additional photography. I would also like to thank my team manager at the Design School Kolding, Lene Nyhus Friis for her professional support while writing this book. Jen Gardner provided eagle-eyed, black-belt copy-editing help and thank you to Grace Harrison for patient editorial support.

1 How do we see?

1.0 INTRODUCTION

The photo in Figure 1.1 is not a tram. It is a set of grey tones that has been interpreted as a tram. This interpretation happens courtesy of visual systems in your brain which are outside your conscious control. Your brain can't help but see what is not there. Your mind has interpreted the shading, the verticals and the horizontals as a three-dimensional space in which other objects are situated. The fact that it is a picture first is something we have to actively bring to our consciousness.

The mind can be understood as an integrated group of mechanisms determined to interpret the world in set ways, following physical and biological laws, overlaid by cultural influences. It can find a moving animal in forest foliage and distinguish between holes and shadows on the ground. And it can see in a flat mass of grey shades an image of something that is not really there. The world we perceive is derived from light falling on a pair of 27-mm spheroids inside your skull. Nobody knows what the world really looks like.

Nature has no pictures and no artificial objects. However, the unconscious systems of the human mind are not aware of this. There is thus a tendency to misinterpret visual data if it is not organised (designed) properly. This is in part an explanation of why we misread artificial objects. It is also a warning to designers that they need to create new shapes that do not trick the mind's visual systems into seeing what is not supposed to be there.

In this chapter I introduce the systems quietly operating below our level of consciousness. People's minds have a strong tendency to see things certain ways. This chapter explains why that is by examining the mechanisms responsible for seeing what is not there or seeing things the wrong way – occurrences labelled as a 'design mistake' or 'bad design'. Bad design is as much a feature of the mind as the object it perceives.

The aim of this chapter is to provide several explanations of how we see, each leading to an aspect of visual form. The central puzzle for designers is why creating new and original shapes is so hard and why it can be so difficult to assess what we see.

DOI: 10.4324/9781003183303-1

1.1
This is not a tram. Titled with due deference to René Magritte.

Designers do two things. One: look at the world around them, including new shapes and forms. Two: make new forms. These new forms have to look acceptable and to be seen by others to look acceptable or, better still, look beautiful. If you have some insight into the complexity of the human visual system, you will understand why getting design right is not easy and why there are competing ways to evaluate designs. If one understands some of the possible ways to explain the human visual system, then the difficulty of the task might seem a little less of a surprise. Furthermore, it will help avoid common pitfalls in process of giving shape to new design ideas.

What this chapter doesn't do is explain the physics and biology of the visual system[1] but rather looks at some of the mental subsystems we depend on to understand the world around us.

Some insight into the complexity of the visual system can be gained by a looking at the diagram of part of the visual cognitive systems of a macaque monkey (Grossberg, 2014) shown in Figure 1.2. It's not at all important to know what each box represents – it's the overall jumble of boxes you should be noticing. Each box represents one of the brain functions which could be considered as being analogous to apps. The diagram shows the path of images landing on the retina (at the bottom of the diagram) as they go through the brain and get processed and understood (at the top of the diagram). Notice that the system has connections which jump levels and which have feedback from upper levels to lower levels. The human visual cognitive system is very likely to be at least as complex and as hard to explain as the macaque's. This diagram is a strong hint that visual perception is not a simple matter. If you had a mental model of the visual system, it probably did not look like that diagram.

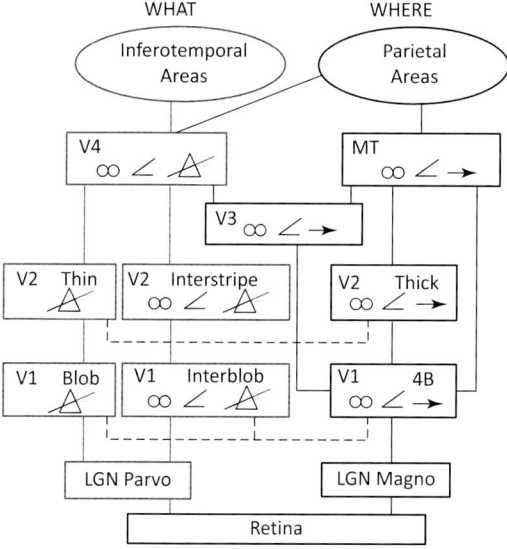

1.2
Schematic model of part of macaque monkey's visual system. The detail is not as important as the apparent complexity of the system (from Grossberg, 2014).

Even if it feels like images are projected from the world through the eye onto a screen in the brain, this not the case. Colour, space, lines and motion are sorted out and combined and related to previous experiences. Further, there is interaction with cultural experiences and cultural values that further complicate perception: the macaque does not see as we do; people from remote tribal communities do not see as we do; people from the 14th century saw things differently to the way we do and the way we view things alters as we develop, mature and age.

This chapter presents six ways of understanding what we see. Gestalt theory explains how objects and groups of objects are perceived. I will present direct and indirect models of visual cognition. These competing (but probably complementary) views are attempts to explain human perception of space and how objects are sensed within it. Anthropomorphism is a phenomenon explained by evolutionary psychological theory, which is especially relevant in the interpretations of objects' friendliness and perceived gender. Theory on primitive forms explains the underlying structures of designed objects in terms of basic geometrical forms. The section on forces and seeing 'as if' borrows from the work of Cheryl Akner-Koler (1994) to provide a way to analyse forms and how viewers interpret forms 'as if' being acted on by forces. There will be examples illustrating how the theory can be seen in action.

Given the complexity of the human visual system, there are a number of possible orderings of this material. Some may prefer Gibson's theory to be explained after Gregory and others insist on the opposite. The section on primitive forms could be placed before or after the section on gestalt. I have chosen to leave it to the end based on its tentative nature.

1. Gestalt theory – explains how we see objects in the environment
2. Gibson's theory – proposes a form of active viewing
3. Gregory's theory – proposes indirect (constructivist) perception
4. Anthropomorphism – explains why we act towards objects as if they were people (or animals)
5. Primitive forms – might explain how the underlying geometry of forms can be perceived
6. Seeing 'as if' – explains how we interpret shapes *as if* forces had acted on them.

BOX 1.1   KEY EXPLANATORY MODELS

Each of the sections in this chapter constitute mere signposts to areas of research which are deep, broad and contested. For example, the work of the psychologist Gibson is complex and its meaning is still being debated. Design researchers are not themselves sure of its implications. Readers are strongly encouraged to follow up the references provided (see the Further Reading list at the end of the chapter). This chapter is not enough to go on. Its sole job is to make the reader aware of the main issues. Colour has been left to another section, and even then only dealt with briefly. This section deals with the perception of what can be thought of as monochrome geometry upon which colour is later draped: dots, lines and volumes.

Figure 1.3 shows how the themes of this chapter are connected.[2]

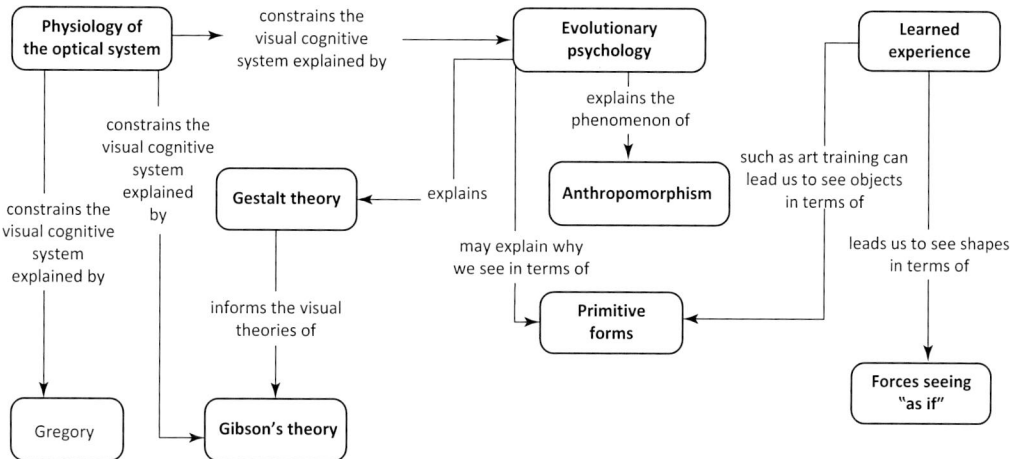

1.3
Overview of the themes of Chapter 1.

1.1 WHY DOES THIS MATTER?

Before dealing with the general principles of perception, I would like to make clear why this theory matters to designers: it is about *why* people see as they do and also *what* they see. With this understanding one can control what the viewer notices when managing form, so the viewer is more likely to see just what one wants them to see. This is what Monö (1997) discusses as 'design for product understanding'. Design errors are often a result of insufficient attention being paid to ways of seeing; either there is too much 'noise' getting in the way of the design intent or what little is there is confusing. Many aesthetic design errors that I see in production or in the classroom can be classed according to the sections of this chapter. Having seen so many down the years I have been able to find out why the errors are errors and to build up a general-level understanding of their nature.

It would be helpful to distinguish between noise and confusion, which are two broad headings of error:

Noise: *'I can't see what this really is'*. Example: an uncoordinated suburban environment; confusing poster graphics; bright lights and wet pavements on a nighttime street; the view through a rain-washed windscreen on a stormy night.
Confusion: *'I see this thing and it doesn't make sense'*. Example: a rounded feature on an object composed of sharper edges; a coloured shape distracting from an overall form; a switch that looks like it should be turned but is really supposed to be pressed.

Interestingly, examples of truly and unambiguously poor design are not easy to come across. Figure 1.4 shows an instance from architecture. Having looked hard to find examples without much success, I realised ugliness is not necessarily the opposite of beautiful or pleasing. Rather, it is disorganised, context dependent and a combination of the intentional and accidental. It might be that the paucity of examples is due to the fact most of the worst ideas do not reach production or are simply not deemed to be worth photographing. Much of what is called 'bad design' (as opposed to ugly) consists purely of poor ergonomics and disappointing functionality. That is indeed a form of bad design but is not to do with aesthetics or appearances only. Indeed, a lot of 'bad' design looks quite alright. It has an acceptable appearance but is hard to use or understand.

1.4
Apartments and bus stop, Aarhus, Denmark. Individually the elements are at least tidy and neat, but as a whole, it is badly designed (or not designed at all). Bad design seems to involve a lack of control or loss of control regarding how the totality of the ensemble is perceived. I had originally picked an ugly Belgian house for this illustration but decided that it was an artistic statement and not bad design at all. The search for ugly or badly designed things led me to realise that ugliness is often about context: things that don't fit in or don't work together. The sheer ghastliness of many urban settings (design that is bad) is often not easily photographed. But being there is demoralising because the place shows no concern for the feelings of residents. It is mere engineering. Architects and planners have been much more prolific than designers in making sad and bad things. At least with consumer goods you can avoid them or dispose of them if they are too nasty.

In this chapter we are interested in approaches that help in organising pure shapes so that they are at the very least inoffensive. It will also help creative, expressive designers organise the form language so that is coherent and consistently applied.

To understand why we see as we do, we need to consider the origins of the visual system. The human cognitive system evolved in response to the need to navigate in a hazardous environment, to find food and to avoid being eaten by predators. This is the same system we now use when gazing at a building, a landscape or a motor car, as well as when looking for food in the supermarket. This system has the added richness of cultural and learned influences.

Sight and visual comprehension are taken for granted, which is why the complexity of the system is often underestimated. One might have the feeling that sight is a passive process, that images fall into one's eyes, that the world is self-evidently 'just as it is'. However, the human cognitive system is made up of a set of interlocking, dynamic modules charged with different functions, and this complexity permits several competing explanations of visual perception. It also explains why perception and aesthetics are subjects of ongoing discussion. Further, an individual's visual perception evolves and matures such that people at age 1, 10, 20, 30 and 40 onwards see the world differently (Sowden et al., 2000; Luna et al., 2004).

Please remember there are two parallel strands to discussing visual cognition. One is the functional approach, which concerns *how* images are seen and processed in terms of light, colour, geometry and motion (Pinker, 1984). This approach is taken by engineers designing image recognition systems, such as those that allow robots to move about in and interact with environments. The second approach concerns aesthetics (Dake, 2004); that is, how we interpret and find meaning in what we see (e.g., Nygaard Folkmann, 2013). It is safe to say that engineers are not designing robots that find deep yellow beautiful or that will admire the forms of an ancient Italian village. These capabilities are not needed for an engineering solution. The aesthetic approach works at the level of the value analysis of images ('that's a truly blue painting!'), the level of meaning ('that curve may refer to Japanese woodwork') and at a level of emotion ('that painting gives me strong feelings I can't even describe!'). While the aim of this book is to help designers create and analyse objects that are aesthetically pleasing in the second sense, I will not focus much here on the philosophy of aesthetics (the long and unresolved attempt to explain what it means when we experience a 'wow!' moment). This chapter's emphasis will be on the main theories of how we understand visual perception. Managing form can lead to a higher likelihood the viewer will experience a 'wow!' moment, an aesthetic experience. The focus here is on explaining how we even see and make sense of the thing that causes the 'wow!' moment.

1.2 VISUAL COGNITION

There are six fundamental approaches to understanding visual cognition in the biological sense (see Box 1.1). The most relevant aspect for designers is that these approaches make us aware of viewers' cognitive biases: such as why viewers tend to see faces in patterns, why sharp-edged and deeply coloured things look closer than rounded pale things and why our senses mislead us about a form. We will start with gestalt theory.

1.2.1 Gestalt theory

An early and useful approach to visual cognition is called gestalt theory, which overlaps with Gibson's work (see Subsection 1.2.2). *Gestalt* is the German word meaning 'the whole object'. Gestalt theory is a body of knowledge concerning shape and motion recognition. It was developed initially by the German psychologists Max Wertheimer, Kurt Koffka and Wolfgang Köhler in 1912 (Behrens, 1998). At a very basic level, gestalt theory can be understood as a set of rules for how we see shapes in relation to the whole assembly in which the shapes are positioned.

> There are wholes, the behaviour of which is not determined by that of their individual elements, but where the part-processes are themselves determined by the intrinsic nature of the whole. It is the hope of Gestalt to determine the nature of such wholes.
>
> (Wertheimer, 1938a, p. 2)

Its most basic application is to describe in general terms how objects are discerned in their environment and what their spatial relations might be (in front of, behind, advancing, receding). Gestalt theory can be understood as the simplest way to interpret an image. In retrospect, gestalt theory seems to be dependent on what is called evolutionary psychology but was formulated before that theory was proposed. We will return to that in Section 1.4.

Each gestalt principle seems to be related to a necessary mechanism in the brain to analyse the images being sent from the eye along the connecting nerves to where it is processed. Consider a gazelle or a human standing about somewhere. Its brain is configured to have systems which analyse the surroundings for defence and to find or avoid things. There are pre-existing structures inside the brain that have evolved over time to do this. They fulfil a group of related functions. The organism will first need to distinguish between an object (the 'figure') and its background (the 'ground') and between moving and non-moving objects. The organism also needs to judge the size of the object, its speed and the direction of motion.

The general principles of gestalt theory are formulated as several laws (Wertheimer, 1938b).

> Figure-ground. This law appeared later in the development of gestalt theory but is listed first as it is quite dominant in its effect. It states that the boundary

line between an object or figure (see Figure 1.5) and the background belongs to the figure. Note that this is relative to the immediate field of view. Compare this to the effect in Figure 1.6.

1.5
Figure-ground. On the left, the edge belongs to the black shape, which is a black figure on a white background. The middle image is a figure on a ground. What is going on with the image on the right, though? The kettle is a figure on a ground *and* the photograph of the kettle is a figure on a ground (the page). Which do you see most strongly?

1.6
Figure-ground. In this instance a white oblong can seen as a hole in a black triangle. The edge now belongs to the black triangle, which is sitting on a white background. You can also actively choose to see the white area as a box on a black background.

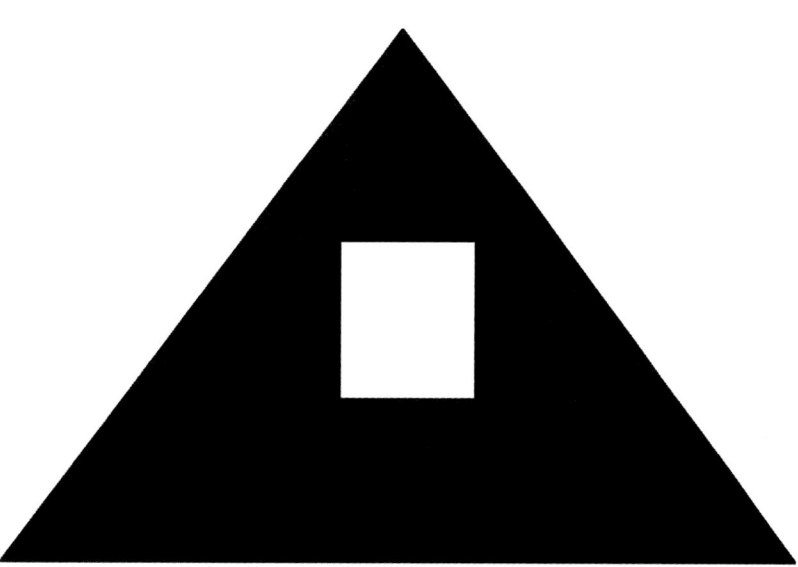

'The Factor of Proximity' law states that objects that are close to one another will be perceived as being related to one another and thus be part of a larger whole, or group. The diagram (Figure 1.7) is seen as showing two distinct groups rather than 24 squares. It is, in fact, 24 squares. The 'groups' are not there but are an artefact of our mental biases.

The photo in Figure 1.8 (the telephone) shows the proximity effect in operation and how subtle it can be. The distance between the 12 central keys, marked 'A',

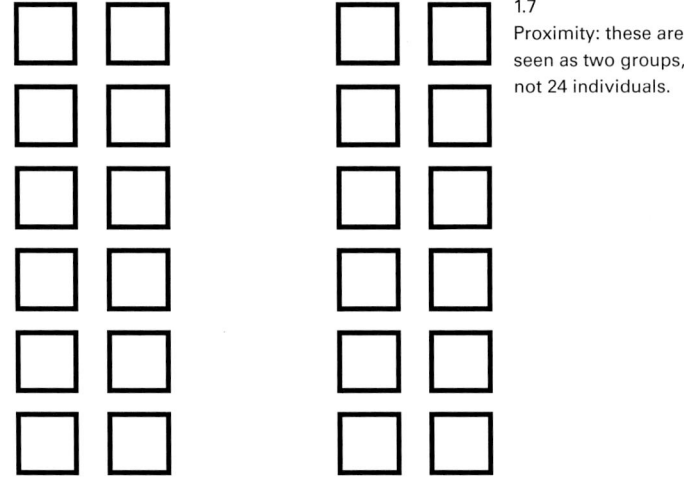

1.7
Proximity: these are seen as two groups, not 24 individuals.

1.8
The area 'A', outlined with a box, is seen as one group. The six other buttons are perceived as two subsidiary groups. The arrangement makes sense given their functions.

and the groups of three buttons left and right of them is only marginally greater than that between the 12 keys.

'The Factor of Similarity' law states that objects that look similar or have similar properties are assumed to be associated. Such properties are size, colour and shape. Figure 1.8 shows identical buttons arranged in a keypad. Figure 1.9 shows it in the abstract. The grey squares form groups distinct from the white ones so that what you tend to see is rows of alternating grey and white rows and not 18 grey squares and 18 white squares or 36 coloured squares.

1.9
Similarity: the grey
squares are seen as
units (rows).

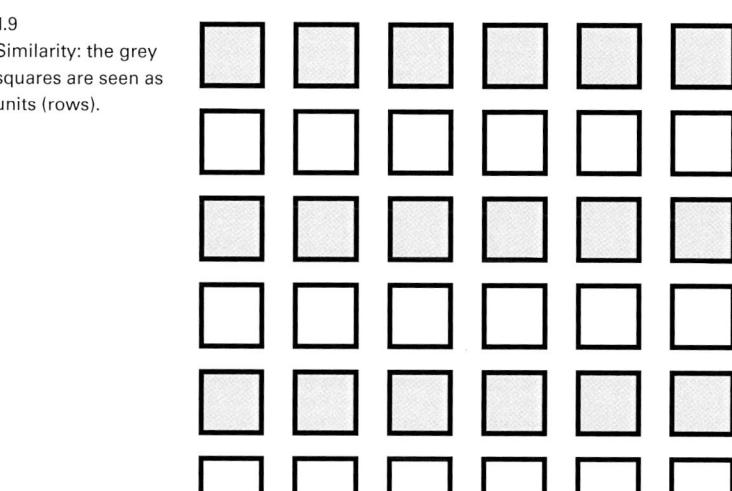

'The Factor of Uniform Destiny' (or 'Common Fate') (Figure 1.10) is the phenomenon experienced when a group of objects are perceived to have a sense of motion or directionality. Whichever principle is in operation generally applies to the individuals. An example would be when a flock of birds is in flight. The movement of the group is assumed to apply to the movement of the individual. A bird moving in the opposite direction would not be perceived as part of the whole.

1.10
Common destiny.
We can see the
bubbles arising from
the pipe as grouped
as they look as if
they are heading
the same way. The
single bubble is not
seen as part of the
group of bubbles.
It looks as if it is
heading in another
direction. The notice-
ability of the one
bubble is marked.

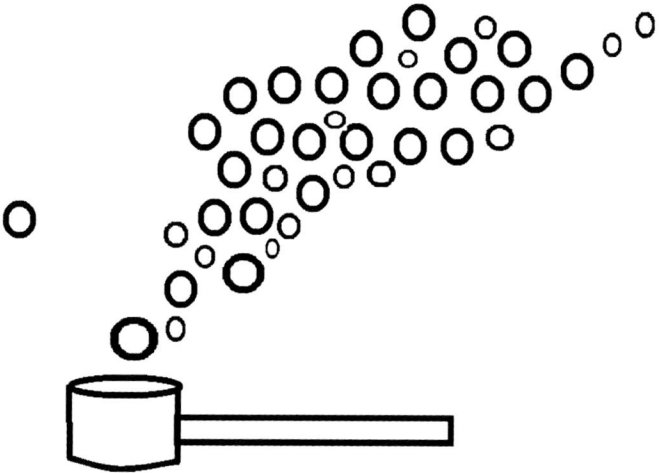

'Prägnanzstufen' (also known as the 'Law of Simplicity') states that we perceive the simplest arrangement of elements first. In a simple case, one sees the lines in Figure 1.11 as five ovoids rather than as 17 curves.

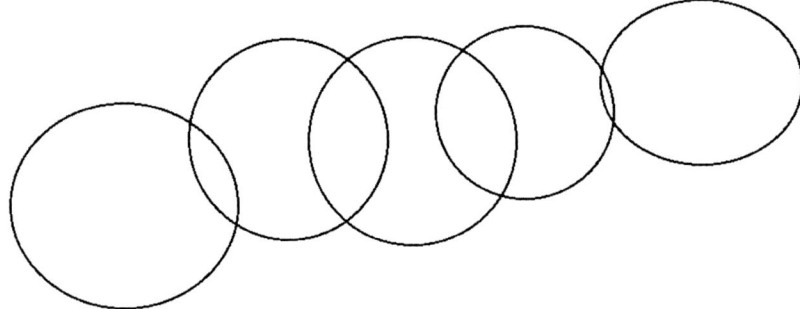

1.11
Law of Simplicity:
these 17 curves are
seen as five ovoids.

In a realistic example, when confronted by an object such as a telephone made up of black squares, a cable and a handset, this is seen as one object rather than as several different items (see Figure 1.12). By 'simple' I mean fewer rather than more items and symmetrical versus asymmetrical.

1.12
This telephone is
seen as one item
even though it is not.

'The Factor of Direction' (or 'the Law of Continuity', or 'Good Continuation') (Figure 1.13). Where the viewer sees a group of lines, the most likely interpretation is that lines that run in the same direction are continuous with each other. This assumes the least number of parts.

'The Factor of Closure' law states that almost-closed curves are interpreted as closed (Figures 1.14–1.17). The mind tends to extrapolate the curves. This tends to produce a simpler interpretation of the shape. Many corporate logos make use of the phenomenon. The World Wide Fund For Nature logo shows what is seen to be a panda. In fact, it is a set of unconnected dark areas interpreted as a panda.

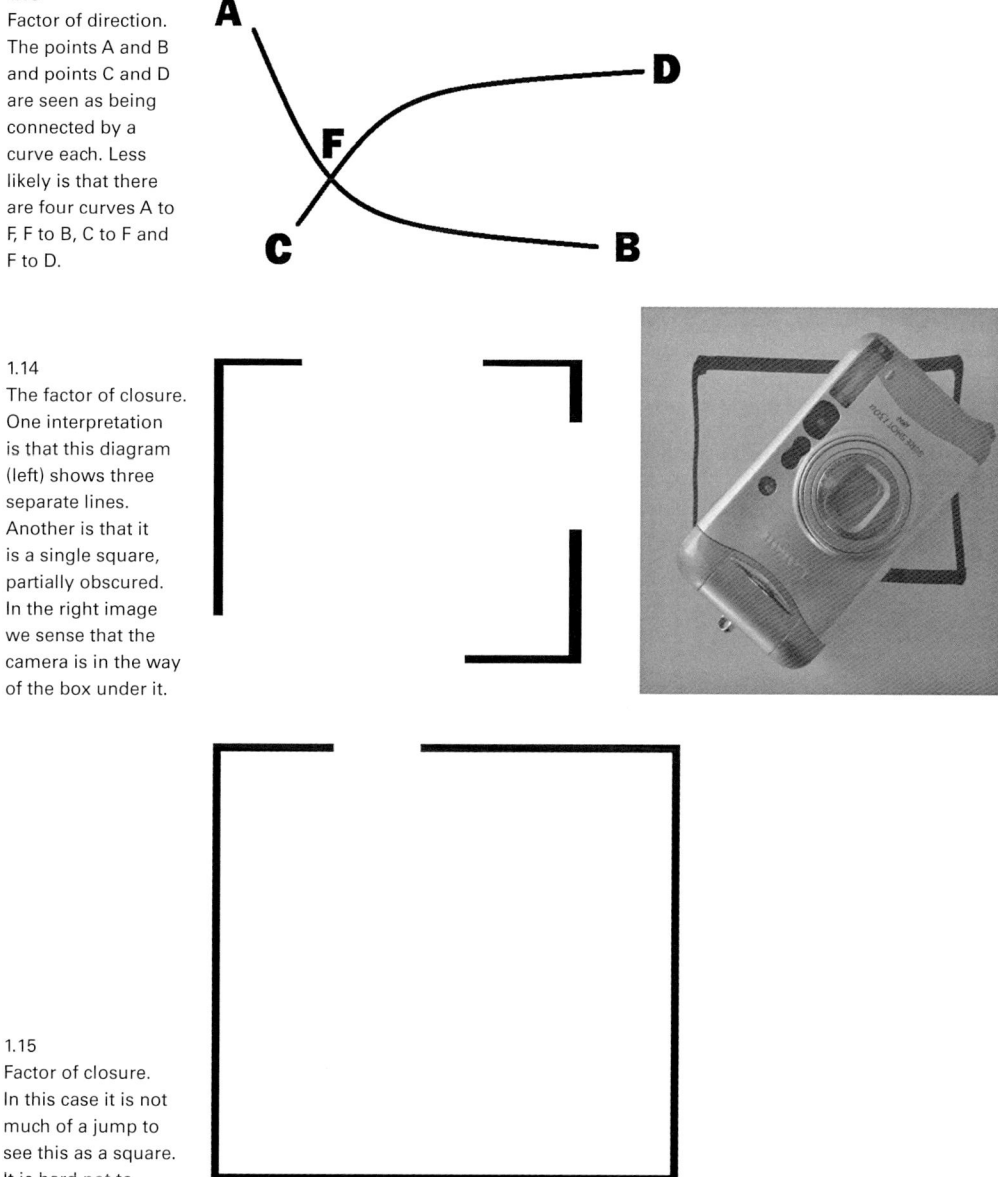

1.13
Factor of direction. The points A and B and points C and D are seen as being connected by a curve each. Less likely is that there are four curves A to F, F to B, C to F and F to D.

1.14
The factor of closure. One interpretation is that this diagram (left) shows three separate lines. Another is that it is a single square, partially obscured. In the right image we sense that the camera is in the way of the box under it.

1.15
Factor of closure. In this case it is not much of a jump to see this as a square. It is hard not to.

'The Factor of the "Good Curve"' (see Figures 1.18 and 1.19) states that if there is a strong sense of linear continuity between two curves, lines or groups this can override other gestalt principles. In the photo, a car's tail lamp could be seen as a figure on the ground of the car's body. There are lines (shown in white) running into the lamp from outside it. In this example, the various principles appear to be in conflict. Are we supposed to see the lamp as an item on its

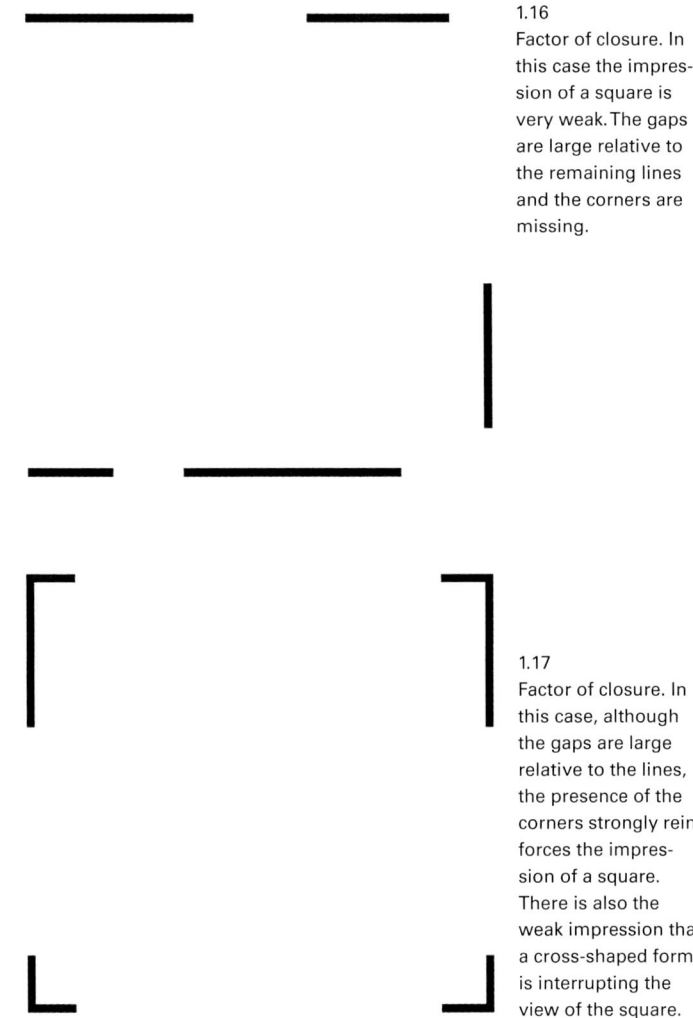

1.16
Factor of closure. In this case the impression of a square is very weak. The gaps are large relative to the remaining lines and the corners are missing.

1.17
Factor of closure. In this case, although the gaps are large relative to the lines, the presence of the corners strongly reinforces the impression of a square. There is also the weak impression that a cross-shaped form is interrupting the view of the square.

own or as something integrated with the surrounding shapes? Since the lamp cuts into the window, it suggests the lamp might be seen as dominant. That the lines run from outside into the lamp suggests that the designer wanted the lamp to be integrated. That another curve crosses the lamp further confuses the effect.

This is a general introduction to gestalt principles. Note that the classic diagrams used to illustrate the effect are two-dimensional. Industrial design operates in three dimensions and includes colour. As such, there are very many circumstances when the laws interact with complex geometry and varying materials. Colour perception has its own set of rules and often colour can dominate other characteristics such as line and surface form, at least initially.

1.18
Factor of good curve? This shows an ambiguous design with several laws in conflict.

1.19
The same car seen from the side. The line AB is suggested by the internal structure of the lamp. The line BC is a compound of the glass edge and a black line in the lamp. The shape D is the outline of the lamp. So, is the lamp a figure on a ground? It is ambiguous.

EXERCISE 1.1

The aim of this exercise is to take a case of the gestalt principles and explore its limits.

- How far apart do lines need to be before they are not seen as an interrupted whole?
- Can you combine two gestalt laws and see which one is dominant? What about combining the law of continuity with the law of good curve?
- Try to find examples of gestalt laws in every day design objects. Control panels seem to be an especially fruitful area for this.
- What effect does colour have on gestalt principles? What would you expect?
- Try taking a principle but experiment with two or more colours to see how they interact.
- Look for designs that might exploit a gestalt law: windows arranged on a building? A case from graphic design? Do product and automotive designers use these effects? Or fail to use them?

The car on the left is the original. Note the line under the arrow. On the right, the line has been removed. Do you think the line was needed in the first place? Is it better without it? (Image: Daniel O'Callaghan.)

1.2.2 Direct perception (Gibson) and indirect perception (Gregory)

Before launching into this section, the reader will want to know immediately what these ideas can be used for. Gibson's ideas on direct perception are the basis of design considerations regarding usability, popularised by Donald Norman (1988), who elaborated the concept of affordances. Anyone who has been confounded by the controls of an electric oven or can't work out how to operate a device will have experienced an object with unclear affordances. Gregory's ideas are more subtle and relate to how we make sense of what we see based on past experiences and also how we can come to accept designs which seem at first to be difficult to like.

What are affordances? In *The Ecological Approach to Visual Perception* Gibson (1979) wrote:

> The affordances of the environment are what it offers the animal, what it provides or furnishes, either for good or ill. The verb to afford is found in the

dictionary, the noun affordance is not. I have made it up. I mean by it something that refers to both the environment and the animal in a way that no existing term does. It implies the complementarity of the animal and the environment.

(p. 127)

According to Maier and Fadel (2007, p. 1), 'an affordance is what one system provides to another system'. Brown and Blessing (2005, p. 3) surmised that 'one could consider the affordances of a device to be the set of all potential human behaviours that the device might allow. This, of course, is a very large set'. How about some concrete examples: a chair affords sitting while a knife affords cutting. However, you can also use a knife as a paper weight or a prop to keep a door open. I once used a fleece jumper as a pillow for a while. The term *affordances* is slippery, but we can constrain it to relate to the most obvious purpose we wish the user to perceive; for example, that the door handle is for gripping. In some cases affordances are not clear and the user is confused.

You may notice that gestalt theory and those of Gibson and Gregory together add up to a quite comprehensive but not fully integrated theory of visual cognition. Gestalt theory does not depend (historically) on a wider view of the visual system, whereas Gibson and Gregory make larger claims. It might help to view Gibson's and Gregory's competing views as being complementary. Whilst cognitive psychologists and design researchers are still debating their relative merits, we can use them as handy rules of thumb.[3]

Figure 1.20 explains how gestalt, Gibson and Gregory relate to one another.

The work of Gibson (1966, 1972) involves what is known as direct perception. That means it assumes that seeing is not mediated by conscious thought

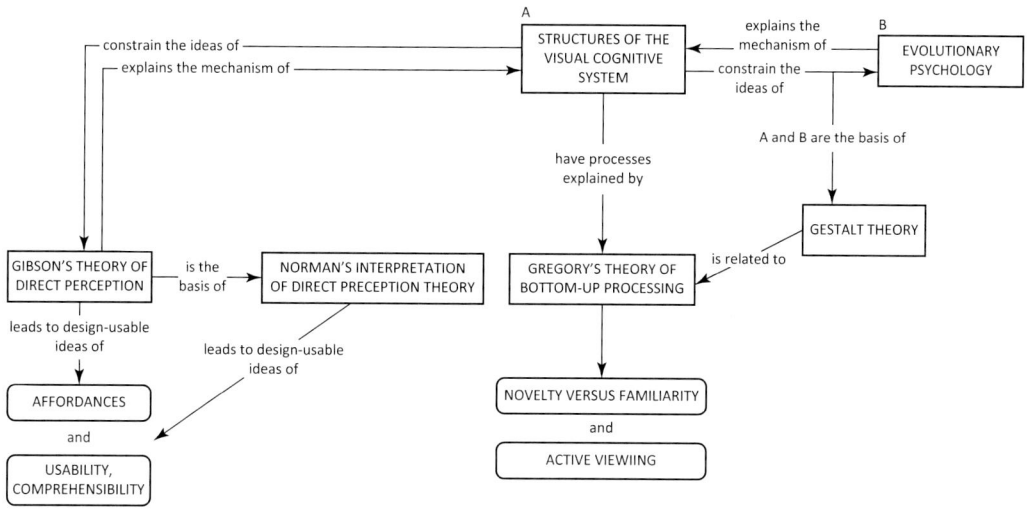

1.20

How gestalt theory and the theories of Gibson and Gregory are related.

processes. Gibson's work was inspired by research into how pilots made sense of the visual data they perceived when landing planes. He posited that there are objects moving in the environment and there are features of those objects that are constant. For example, the actual size of the runway is a constant but the change in its apparent size informs the viewer how far away one is and how quickly they are approaching it. The relation of the observer and the environment, and other affordances, plus innate cognitive structures allow the viewer to make sense of what they see. From a designer's point of view this means people have certain biases of which the designer must be aware. For example, lines connecting two points are taken to mean a connection of some sort. To graphically link two functionally unrelated controls could lead to user error. Although the way in which we perceive the affordances of objects is contested, the concept has been adopted by design theorists such as Donald Norman (who has adapted Gibson's ideas).

Gibson's approach suggests that we interpret visual stimuli arising from our surroundings and that context affects the way we do so:

> When the constant properties of constant objects are perceived (the shape, size, colour, texture, composition, motion, animation, and position relative to other objects), the observer can go on to detect their affordances. I have coined this word as a substitute for values, a term which carries an old burden of philosophical meaning. I mean simply what things furnish, for good or ill. What they afford the observer, after all, depends on their properties.
>
> (Gibson, 1966, p. 285)

The brain takes in the incoming information without much interpretation. Elements of the context (the affordances) that support the correct analysis of what we are seeing include the relation of the object to the horizon; the compression of texture, which implies distance; and whether ladder patterns are emerging from or converging on a point (Figure 1.21). The way objects further

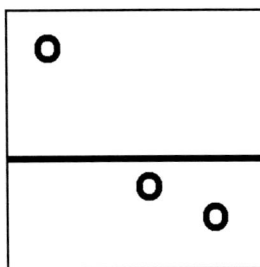  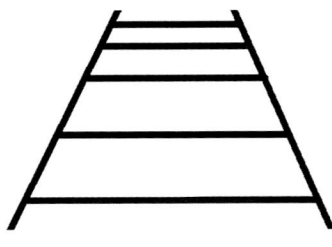

1.21
In the left diagram the three circles are interpreted as having different sizes. The lowest one is seen as smallest. The top left one is the largest and is understood as furthest away. In the middle diagram, the compression of lines is seen as showing increasing distance. The right diagram is interpreted as showing parallel lines receding. The lines are not actually parallel.

away are lighter in tone is another clue to relative distance. Painters make use of this when colouring distant mountains light grey, to suggest the diffusion of light by particles in the air.

The significance for designers of Gibson's theory relates both to two- and three-dimensional design. At a general level, the designed object's form must support its intended function and the object may inform the user's behaviour:

> an organism and its environs are reciprocally shaped; perceptual features are adaptively moulded in response to specific environmental features; both simple and complex organisms exhibit patterns of response to stimuli that are demonstrably innate.
>
> (Jenkins, 2008, p. 44)

Thus, when confronted by a new and unfamiliar object, the viewer may make an error regarding the affordance. The designer is responsible for how easy or hard this might be to do. The relation of this point to the aesthetics of design is that worthwhile artistic, expressive goals may sometimes be in conflict with the need to spell out what the object does. A further way to use Gibson's idea of holistic, direct perception is as a counterweight to the tendency to look at objects in detail only. A common problem in design is to see features in isolation only. Designers should also try to look at their work as a whole, to try and take in the object as one entity.

A critique of the concept of affordances was provided by Oliver (2005) and a critique of Gibson's model of perception can be found in Ware (2012). Specifically, Oliver was sceptical about Gibson's claim that no mental activity is involved when the observer sees objects in the environment. Ware raised questions about Gibson's work in relation to graphical user interfaces. Ware concluded that 'we can be inspired by affordance theory to produce good designs but that we cannot expect much help from that theory in building an applied science of visualisation' (Ware, 2012, p. 20).

EXERCISE 1.2

That aspect of Gibson's work that is most useful relates to affordances. Take a look at some objects around you and examine their affordances as in the most obvious thing the object can allow you do.

Notice that the affordance is expressible as a verb: chairs afford *sitting*, knives afford *cutting*, a door handle affords *gripping*.

Next, choose an affordance and draw an object that can provide it; for example, a button affords pushing. Then redraw it to confound the affordance. Experiment with the degree to which this is apparent. The point

---

here is to see how robust the affordance is or how frail the affordance is. Design failure often occurs where the object suggests either a false action (it suggests pulling but you need to twist) or an ambiguous action (should it be pushed or pulled?).

You could consider safety measures as anti-affordances. A childproof bottle top does not (easily) afford opening. How could you make a safety chair? Public benches with metal dividers do not afford lying on them and so deter homeless people. Such design is ethically questionable.

Design for disability is related to maximising the affordance of an environment: level surfaces and reachable switches. Is your environment designed to include people of various physical abilities?

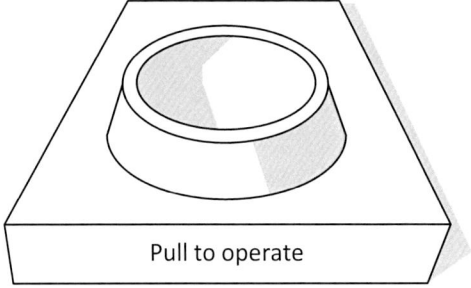

Pull to operate

What is wrong with the control illustrated in this figure?[4]

Now we turn to an opposing theory of perception, Gregory's (1970) indirect model of perception. If you have ever wondered why you get used to new designs such as initially strange fashions or awkward-seeming cars or incongruous buildings, you have encountered something Gregory can explain. The lesson is that users can get used to new and striking designs if given some time and exposure.

Gregory proposed that through experience we learn how to makes sense of the visual field. His is an indirect model of perception, with cognitive processes informing the assessment of the image such that perception is a form of hypothesis. Again, this theory draws our attention to the biases viewers are prone to. A more specific example of context-dependence is that it is easier to interpret a hand-written word when it appears among other words on a page. Another example is that if a green, round, shiny shape resembles an apple (and there are other apples in the field of view), there is a high likelihood the mystery object is also an apple. Gregory considers this phenomenon to be a form of reasoning analogous to scientific practice (but running subconsciously). It is not as odd as it sounds. As we get more information, we begin to consider the possibility that the object, in this case, is indeed an apple. In a sense, we develop a theory or hypothesis that the thing may be an apple. As more data come in, we either conclude that it is an apple or decide that it's something else like a tennis ball.

To show that this reasoning is analogous to scientific reasoning, the so-called hypothetico-deductive process, I have made a table comparing the steps (Table 1.1). They occur at different speeds and involve data of different complexity but are similarly structured.

Table 1.1 Comparison of the rapid process of indirect perception with hypothetico-deductive reasoning.

	Indirect perception	Hypothetico-deductive reasoning
Step 1	Sense: there's something there	Observation: notice a phenomenon
Step 2	Wonder if it's an X	Form a theory to explain X
Step 3	If it's an X I must check if it has some feature I can look for …	If the theory is correct, then I should expect Y to be the case
Step 4	Look a bit closer, wait for a better view, turn on the lights, etc., look for the feature	Actively get data to see if Y is the case (such as an experiment)
Step 5	Take in new information	Assess data

According to Gregory, much of the information that lands on the retina is eventually discarded by the time it reaches the point where the conscious mind makes sense of it. This has direct relevance to designers. First, the more information the brain is confronted with, the more it has to process. In discarding part of the signal, some valuable information may be lost. Or, the less there is to see, the easier it is to interpret. It is easier to find a sock on the floor than a sock among lots of other socks (see Figure 1.22).

Second, the design must be organised so that when the surplus information is discarded, the remaining information makes sense based on the likely past experience of the viewer. The ticket machine in Figure 1.23 is not helpfully organised and is visually overloaded.

The case supporting Gregory's hypothesis is as follows. First, unusual and strange objects tend to be mistaken for more familiar objects. This is becase we draw on past experience. Gregory demonstrated this by showing how a hollow mask of a face can be viewed 'incorrectly' (see Figure 1.24). Under certain conditions it is not possible to tell if you are looking at the inside of a mask (the concave view) or the outside of the mask (the convex view). When a mask is mounted on a spinning axle, the viewer can see the actual inside as the apparent outside until the mask rotates enough so that this view is not tenable.

Further, drawing on experience, unusual and strange designs may look unsettling precisely because they can't be reconciled with familiar objects. Designers face the conflict between the need for novelty and the need for recognisability. Interesting designs balance the two.

1.22
A box of socks.

Another support for Gregory is that perception allows us to respond correctly to other characteristics we have not, in fact, sensed in the moment. We learn that oblong areas with light passing through are usually the signs of a partially opened door. So if we see a new instance of an oblong with light passing through it is highly likely it is a door and so we can act appropriately. For designers, this means being careful with references to existing objects. If the design provides the signs that a thing is, say, a moving part, then it had really be a moving part; for example, a switch. From a positive point of view, a designer can provide visual signals of a what an object is by referring to commonly perceived characteristics of the class of thing in question. Or the designer can entertain by deliberately confounding expectations.

1.23
There is a lot to take in here. It's a ticket machine with a lot of surplus information (in Gregory's terms) and the affordances aren't clear (Gibson's terms).

Another supporting argument for Gregory is that one can actively change the way one sees things. The image is the same but the way of seeing it can be chosen by the viewer. The Necker cube (Figure 1.25) is a classic case. In this instance, the gestalt principles of continuity and convergence are in conflict. The diagonal lines suggest distance, but it is not clear which of the vertical lines are in the foreground or background.

According to Gregory's theory, the Necker cube allows two equally plausible explanation for the image. One is that the upper square is nearer and the other is that the lower square is nearer. This creates an unstable image which the viewer can actively see one way or another. This alerts us to the idea of

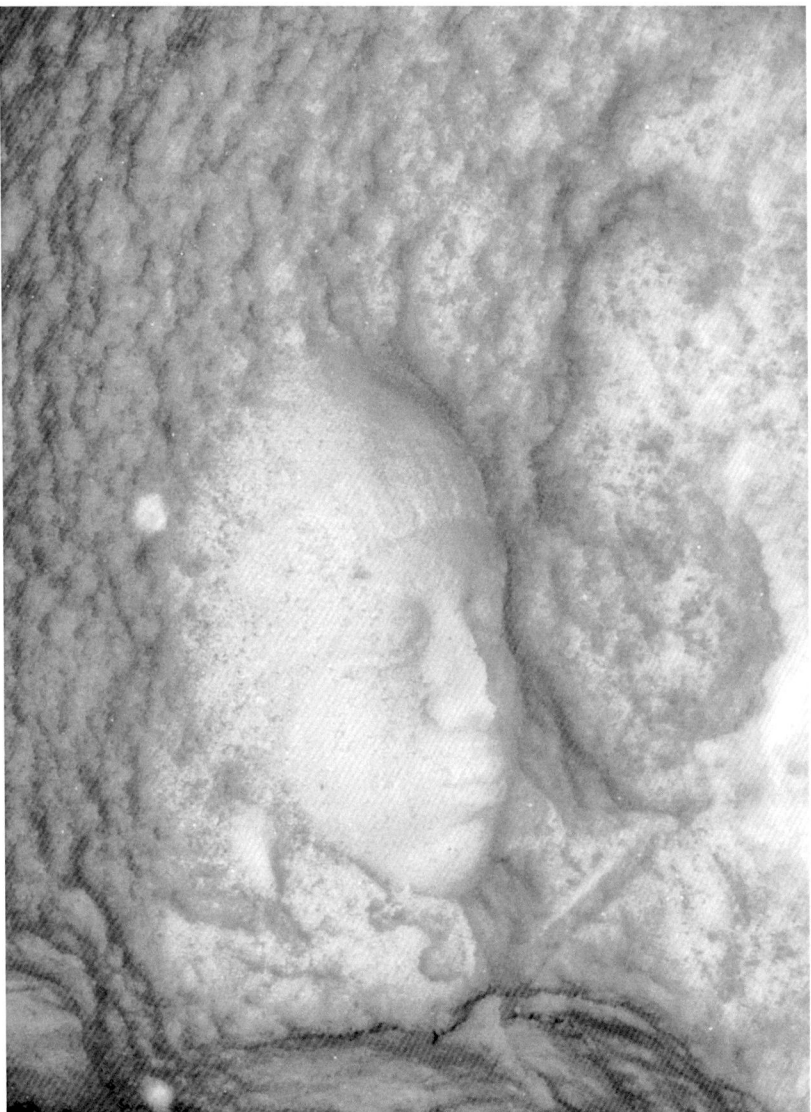

1.24
This image show a similar effect to the spinning hollow mask phenomenon. It shows the impression of a face in snow. It can be seen as a hollow form (concave) or as a form projecting out from the snow (convex). It depends on what you want to see: what you know is there and what you think is there. (Image: Wiki commons)

active seeing: that the mind is to some extent in charge of how to consider what is in view.

Gregory's argument is not immune from criticism. Gregory describes perception as being based on hypotheses. Based on previous experience, the viewer might reckon (hypothesise) that if they see certain phenomena associated previously with X, then they guess they are seeing another instance of X. So, if they see green, shiny spherical objects which are the features suggesting apples, then the viewer might assume it's another instance of real apples.

1.25
Necker cube. Which
square is nearer?

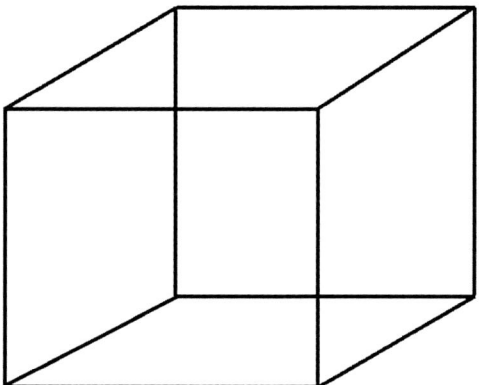

It could, however, be an instance of wax apples or glass apples. Whether we actually have such hypotheses is open to question. The question that can be asked is what sorts of hypotheses these are and whether it is correct to even say that the unconscious mind has 'hypotheses', which are verbal formulations, after all (see Table 1.1). I might hypothesise about my boss's motives, but when I see green, shiny spheres in the fruit shop, do I really hypothesise that they are apples? Recognition seems to be a fairly instantaneous experience. I may not even be thinking of apples when in search of them in the supermarket, yet I am able to find them while occupied by more interesting thoughts.

The counter-argument is to say that Gregory is only speaking metaphorically. Further, our own empirical experience supports the idea that we very successfully correlate partial visual information with actual interpretations. It is very hard to create an object in the world that can't be comprehended. This is why camouflage is seldom 100% effective or effective for very long.

A second argument against Gregory is based on the seeming-paradox that even if we all construct our own perceptions of reality, they are all much the same. Why aren't they very different, since we are all unique individuals? In fact, having a very different perception of reality from others is often a symptom of psychiatric disorder. Another question is this: surely, critics ask, if we had to make up our own different perceptual models of the world, it would lead to a lot of errors and hazards. The response to this is that people experience the one single objective reality 'out there', which limits the options for interpretation. We mostly share similar experiences and so construct similar realities. Second, a more subtle way of expressing the idea that we each have our personal interpretations is that people often notice different aspects of the same thing or don't notice them at all. So, almost all people will judge a form to be a case of X (a portable typewriter, say) but will disagree on whether it is too blue or the right shade of blue, on whether it is too sharp-edged or not sharp-edged enough. Second, natural selection has weeded out the sentient organisms whose visual cognitive skills were unable to 'decode' their environment. Making up your own perception is risky, but each of us is a descendant of an animal that was good enough at the task to pass on the genes for this capacity.

The implication for designers is this: that viewers can become accustomed to new things through repeated exposure. This is indeed why the assessment of new designs changes over time. It is also why the design work of 1900 seems so unlike that of 2020: we are much less exposed to the detailed, hand-crafted work of 1900, so we find it hard to assess. According to Gregory's model, there is too much information in older designs to process easily. In 1900 people were used to busy shapes (and there was a lot less stuff to look at, anyway).

Another critique of Gregory's work is that it relies too much on laboratory experiments. The optical illusions used are only encountered in books on visual cognition. It is true that one does not encounter Necker cubes (Figure 1.25) or Penrose triangles (Figure 1.26) on a daily basis. However, their role is to make apparent the operation of cognitive processes that generally run smoothly and unobtrusively. An experimental device such as the Necker cube makes clear that the mind is set up for a reality where lines that cross over one another usually correclty indicate difference in distance.

1.26
Penrose triangle.

The Penrose triangle demonstrates the strong bias towards seeing pairs of lines as demonstrative of a third dimension and of increasing distance. That Gregory was able to devise instances of how the visual cognitive system could malfunction implies there is a system approximately operating in the way he describes. The tendency for converging lines to be seen as a receding pair is also the basis of the effectiveness of perspective drawing. It doesn't take much to suggest space on a two-dimensional surface (see Figure 1.27).

1.27
A simple drawing suggesting distance and space.

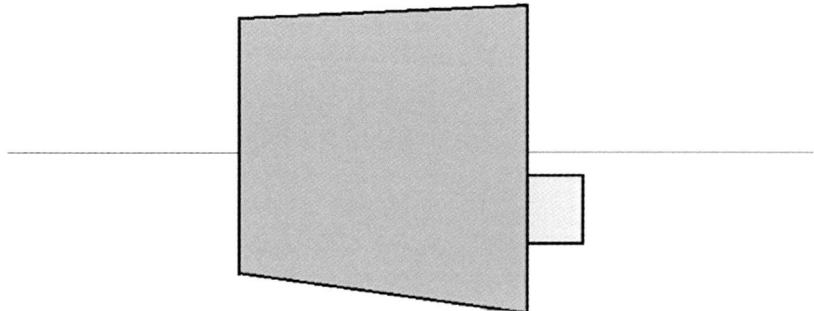

Far fom being contrived and rare phenomena, the optical illusions predicted by Gregory's model are fundamental to representative art and the the reason we see photographs as possessing depth that does not really exist (see Figure 1.1).

There is the question of how the mind makes sense of visual stimuli. Gregory posited that the mind interprets the mass of visual stimuli entering the eye. How does it do this from the moment of birth, at which point the mind has not experienced much visual stimulation? The answer to this is that there are pre-existing cognitive structures in the brain which are primed to deal with colour, light/shade and motion. There are also structures which can arrange the information into three-dimensional understandings such as how changes in light and shade signal movement. What happens in the course of a human's visual cognitive development is that the in-built structures get more and more effective at sorting out visual stimuli, and that is where the hypothesis testing comes in. Gregory is correct to say there is hypothesis testing but underplays the role of innate visual-cognitive structures in the brain. The relevance of this last point for a designer is that over time one's visual judgement alters. One becomes better at seeing objectively one's own work and the work of others.

We need to be careful about assumptions about simplifying designs to make concepts clear. Gibson's work suggests we should make affordances apparent. Gregory might point to designing for clarity and reducing ambiguity. Nygaard Folkmann (2013) proposed that part of what makes for an aesthetically engaging object is what he termed a 'surplus of information'. Information in this instance means the amount of detail or number and complexity of features. A disposable plastic cup has less information than a piece of Georgian tableware. It is in the excess of information of a richly designed object that we find aesthetic interest (when it is viewed in isolation). It is in the ambiguity allowed by the excess of information that we can find different meanings such as the expressive quality of a curve or the details of a classical building. Strict adherence to the the goal of reducing information would possibly result in tedious and banal design. Some designers manage a high level of discipline despite a minimum of information content. The work of Gerd A. Müller for Braun (Figure 1.28) is an example. However, often simplicity can be uninteresting not to say unappealing when it becomes the norm.

1.28
KM3 Food Processor (1957) Mixer by Gerd A. Müller for Braun. (Image courtesy of MAKK – Museum für Angewandte Kunst Köln, Germany.)

As with most competing ways to explain phenomena, there is much that is useful and true in both Gibson's and Gregory's concepts, but there are also explanatory gaps, and perhaps both mechanisms are involved in perception but running in different neural pathways. Humans do have in-built structures to help organize vision (Grill-Spector and Malach, 2004), which is why things 'out there' seem so immediate. But we also control how we see, as in the flip-flopping hollow mask or getting accustomed to novel designs. Conceivably, evolutionary processes have endowed the organism (the human) with generalized systems built into the brain but also with a capacity for adaptation to specific and unknown circumstances. So the question is, which aspect are we designing for?

EXERCISE 1.3

The aim of this activity is to play with affordances by doing the unexpected and 'wrong'. This is learning by deliberately making mistakes.

One of the most significant elements of Gibson's theory is that of affordances. A person will assess the form of an object in relation to what it might be able to do. A kettle affords holding and pouring. A chair affords sitting. Try sketching an object where the affordance is not as one would expect. Try sketching an object where the affordance is plainly wrong, such an as over-styled chair. Can you think of actual examples of chairs where the aesthetic effect has dominated over affordance?

Gregory proposed that the human brain sorts through the incoming stream of stimuli in the surroundings to extract simpler and more intelligible information (e.g., the search for a gun hidden among socks; see Figure 1.22). Look for examples of design that offers too much information in relation to the intended purpose of the object. Can you think of a reason why more than the bare minimum of design might be desirable?

EXERCISE 1.4

The aim of this is to explore the minimum and maximum of style and develop an eye for small differences and the boundary of enough style and too much style.

Find a household object or building or garment. Make a note of its styling features. In one set of sketches, gradually reduce the amount of styling until the object is as plain as you can make it. In another set of sketches, exaggerate the features without adding more. If possible, try to do this as a class. If you swap the drawings with classmates (having mixed the sequence), ask them to rearrange the drawings in the order of complexity. Notice that the original drawing will probably appear in the middle of the sequence. What does this demonstrate? More interesting will be the cases where the original sequence is not recovered during sorting. What does it mean if the drawing of the actual design is located nearer the ends of the sequence?

1.3 ANTHROPOMORPHISM

This section deals with a way of seeing related to the phenomenon of anthropomorphism. The image in Figure 1.29 is automatically viewed as a face even though it's not. It is a set of ovals and curves. Why do we see it that way? Can you even try to see it as ovals and curves? It seems to have an expression. Something powerful is going on when two curves and three ovals looks like a face.

1.29
This isn't a face.

Anthropomorphism is the propensity people have to attribute human qualities to animals and objects. Anthropomorphic effects are used widely in design (see Figure 1.30). Products that make use of it are widespread: vacuum cleaners made to resemble faces, advertisements that use human-like breakfast cereal characters and animal cartoon characters with human behaviours. When we regard a teddy bear fondly it is because of the way its appearance appeals to our instinct to care for babies. The way people view certain cars as cute is due to the large, round lamps that resemble eyes and the soft forms of the bodywork. Anthropomorphism is not only a matter of styling or embellishment but that it can be used as a means of solving design problems. Waytz et al. (2010, p. 219) noted that anthropomorphism might be seen as 'more cute than critical' but that it has pervasive effects on how objects are understood. It is also a phenomenon one might want to consciously *avoid* in a design. Whether you want to use its effects or avoid them, it is important to know what they are and how common they are.

The concept of anthropomorphism in turn rests on a field of study known as evolutionary psychology. For researchers in evolutionary psychology:

1.30
An example of a car design exploiting anthropomorphic effects. This car can be viewed as being cute because of its large, round lamps and smiling grille.

the mind is a set of information-processing machines that were designed by natural selection to solve adaptive problems faced by our hunter-gatherer ancestors. This way of thinking about the brain, mind, and behavior is changing how scientists approach old topics, and opening up new ones.

(Cosmides and Tooby, 1997, p. 1)

Evolutionary psychology is a way of thinking about human behaviour based on the assumption that natural selection affected the evolution of the human brain. In plain terms, organisms that met their environment's challenges better were more likely to survive and pass on the traits that aided survival. Included in the concept of the environment is other organisms. Natural selection operates on behaviour as well as the body of an organism. With this in mind, we can understand how natural selection acted so that organisms (in this case humans) that were better able to interact with others had a better chance of survival. Among those means of successful interaction was the ability to identify and understand body-language and facial expression.

Before moving on, I would like to note that a comprehensive critique of evolutionary psychology was provided by Confer et al. (2010). There are aspects of evolutionary psychology that are disputed and the reference to it here is not meant to be a wholesale endorsement of the field. The disputed areas include how testable the theory is and how much the mechanisms operate in novel environments such as the modern human habitat. Further, how much cultural values play a part is not determined. Evolutionally psychology posits that the human mind is largely the product of the character of the hunter-gatherer period and conditions. Those are radically unlike the ones we experience now. They may not be as important as proposed in the evolutionary psychology theory. For our purposes, we can treat evolutionary psychological reasoning as a useful metaphor.

Turning back to anthropomorphism, a more precise definition is that it

> entails attributing human-like properties, characteristics or mental states to real or imagined non-human agents and objects. According to the familiarity hypothesis individuals draw anthropomorphic inferences, because it allows them to explain things they do not understand in terms that they do understand – and what we understand best is ourselves as human beings.
>
> (Hegel et al., 2010, p. 108)

How does it work? Epley et al. (2007) proposed a three-factor theory of anthropomorphism:

- First, there is the need for an organism to deal with its environment: viewing objects as if they have human characteristics and motivations increases the observer's likelihood of understanding them and what they do.
- Second is that people have an urge to develop social connections and they transpose this to objects displaying human-like attributes. Being kind to toy bears is one manifestation we all recognize. On one level we are aware that the object is not human, and on another we have an emotional sense which means, for instance, punching a teddy bear feels wrong.
- Thirdly, humans experience knowledge about other humans from birth. By induction, they have a tendency to refer to this experience even when judging inanimate objects. Thus, humans learn that a certain expression indicates anger and may view an object showing similar characteristics as having an angry quality.

Anthropomorphism has a further dimension, which is that it leads us also to ascribe gender to objects if there are cues to support this. A chair might look feminine or a power drill might look masculine. This phenomenon is not to be conflated with the sometimes differing preferences of males and females when choosing a product.

Consider the two Danish design classic chairs in Figure 1.31.

1.31
Ox chair (left,
designed by Hans
J. Wegner and
manufacturer
Erik Jørgensen,
Fredericia Furniture,
Denmark) and Swan
(right, designed by
Arne Jacobsen).
(Image: Wiki
Commons)

There is a very high likelihood that many observers would describe the Ox chair (Figure 1.31, left) as masculine and the Swan chair (Figure 1.31, right) as feminine. Starting with the premise that humans are accustomed by either nature or nurture to be sensitive to signs of gender, then they are also prone to reading objects in a similar way. As I always underline when lecturing on this topic, this is not a normative statement but a descriptive one. I do not mean that people *should* see objects in a gendered way. I mean that there is a widespread tendency for people to do this. Where this affects designers is that they need to be aware of the tendency to see objects as having a masculine, feminine or neutral gender and control for it.

EXERCISE 1.5

What this exercise is about is to explore how gender is expressed (or not) in design. There is a lot of room for discussion in this task and perhaps lively disagreement. The aim is not to entrench gender norms but to understand and work with them actively and as critically as you wish.

1. Choose a familiar object and redraw it so that it is more feminine than it appears. Repeat the exercise but increase the masculinity. If you are working with a class, swap drawings and ask your colleagues to arrange the drawings in order of feminine to masculine or vice versa. This task should hone your awareness of the small cues that can suggest gender.
2. Using colour as well as geometry, redesign the chosen object with *mixed* gender signals. What is the effect of this? Can it be used as method to induce creativity? Does it make the design hard to interpret? Ask your colleagues about the resultant drawings. Remember it is possible to have two or three interpretations among the group, none of which are 'wrong'.

3. It might be that very small changes in detailing can radically alter the perception of an object and that the object can be readily 'flipped' from one gender to the other.
4. Check your work with people who are not designers. Find some individuals and ask them to discuss whether they can see gender cues in the design. You may notice that there are two or three competing ways to view the same object. There are no right answers (but sometimes users can give misleading ones!).

The explanation for the tendency to see objects as having a gender has an evolutionary psychological explanation. This can be treated either as a scientific hypothesis or as a metaphor. Either way, the 'story' is that humans are programmed to be sensitive to gender differences because appearances and behaviour are somewhat correlated. Knowing how to relate to another individual depends on having a reliable understanding of their gender. What might be appropriate behaviour between males might not be appropriate between females in the context of group hierarchies and social relations. That explains why humans are so sensitive to signs of gender even in non-human instances.

This sensitivity (as with the sensitivity to babies) is carried over unconsciously to other domains. In design it leads us sometimes to get the feeling that an object has a feminine or masculine character or none at all. This means that the designer should be aware when form-giving that some viewers may see masculinity or femininity in an object. This might have no effect at all (e.g., both chairs are equally attractive and comfortable) or it might not be in accord with the product's expectations. This touches somewhat on the issue of product semantics (i.e., the signs of whether a product is for men or for women or gender neutral; see Chapter 5). The designer thus must consciously manage the object's form so that the meaning or interpretation that is wished for has a better chance of being perceived. The designer has less chance of this if she or he is not alert to the possibility that the user/viewer will be affected by the gender sensitivities.

The flowchart in Figure 1.32 shows the possibilities of action for a designer.

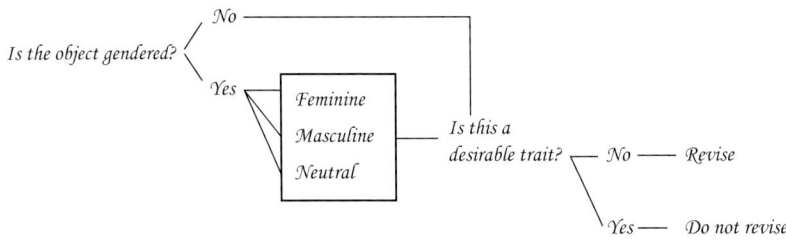

1.32
Design action in relation to an object's perceived gender.

The matter of how the gender of an object is perceived is extremely difficult to judge. Relying on a designer's intuition would be unwise, especially in the realm of consumer goods. Let us assume the designer wants to make the product attractive to females. Then it might seem like a good idea to give the device feminine forms. However, some customers may understandably view this as patronising. One might also assume that overtly masculine forms might appeal to men but in reality customers could find the forms crude or 'macho'. An example might be a certain type of traditionally styled motorcycle which has a very strong appeal to a small number of male customers but might be very unappealing to most prospective motorcyclists. It is a marketing and ethical consideration as to whether strongly appealing to one group is worth alienating a larger group. Further, relying on typical gender signals can lead to clichéd design. And there is an ethical argument that, for example, relying on gender signals in toy design simply imposes gender norms on children that are not appropriate or healthy. For sports products, a lazy approach to design is to assume the male version is the norm and the start point and then to 'shrink it and pink it' for female users. That means reducing the size and using colours associated with femininity. The proper approach is to deal with each user group separately and test their needs and responses.

In conclusion, this section's main purpose is to make the reader aware why people might view inanimate objects as having human qualities and as possibly having a distinct gender. Most real-world cases are more subtle and less obvious than the car or armchairs shown above. The gendered aspect of the design is also only one part of the other aspects of the form. And note that colour can work strongly to distract to alter perceptions of gender: an Ox chair upholstered in hot pink leather will be read differently than the one in tan or black despite the geometry being identical.

EXERCISE 1.6

Consider the font used in Figure 1.32. Is it gendered? If so, what contributes to the effect? What other character has the font?

Some authors point to the appeal of anthropomorphic forms which are indeed widespread. Classic 'cool' industrial design such as that of Dieter Rams is seemingly entirely devoid of human character yet it is also widely respected and admired. Can you think why both approaches to form-giving have been successful?

1.4 PRIMITIVE FORMS

If you have taken a drawing course, you will be aware that a way to structure a new picture is to sketch out the main volumes of the objects. The space within the picture can be managed by treating it as a large rectangle with one, two or three vanishing points. The objects within that conform to those vanishing points and can themselves be blocked out as cubes, rectangles, circles, cylinders or pyramidal forms (such shapes are known as 'primitives'). In this way, a quite complex objects in space may be represented in ever increasing detail. The more experience one has at drawing, the less one needs to actually draw the framing boxes (see Figure 1.33).

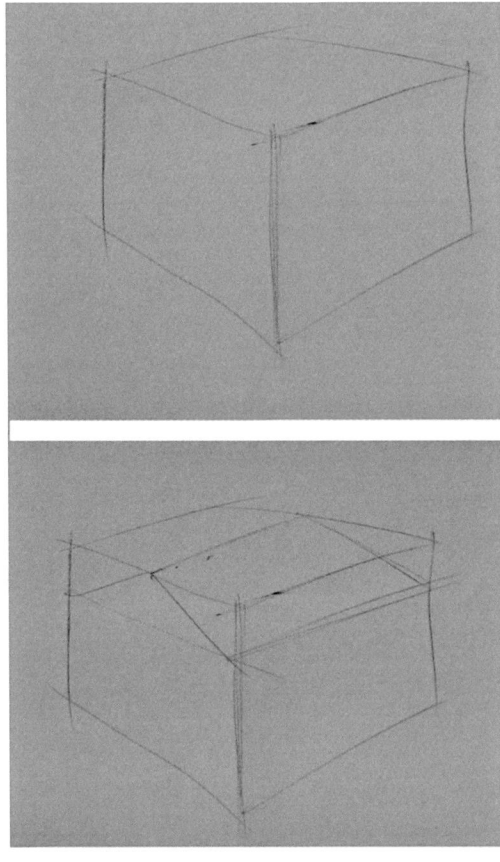

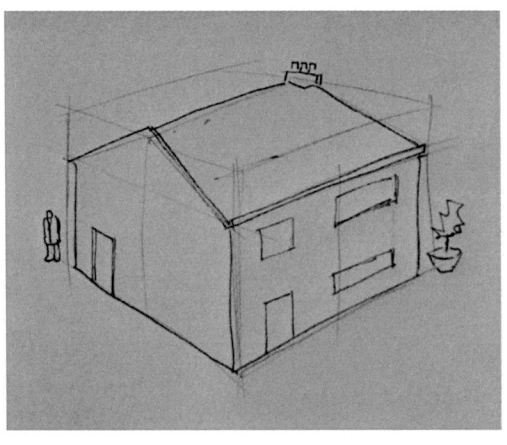

1.33
This shows a house sketched from a primitive form. Our visual understanding of the house may be the reverse whereby we first perceive an oblong and then the details.

Taking an evolutionary psychological approach, it may be that the human mind is so constituted to be able to pick out basic forms amidst a complex field of view. The tendency of newborn babies to react positively to T-shapes means they quickly learn to look for faces, and especially their parents' faces. The T-shape summarises the important geometry of the face and is a form of primitive. Similarly, humans seem to see things in terms of simple basic shapes – a house is a block (Figure 1.33), a door is a rectangle and a person may be a simple stick figure. Consider the image of a car (Figure 1.34) by a six-year-old – it shows the essential attributes of a vehicle, namely, the circles or wheels and an arc for the body.

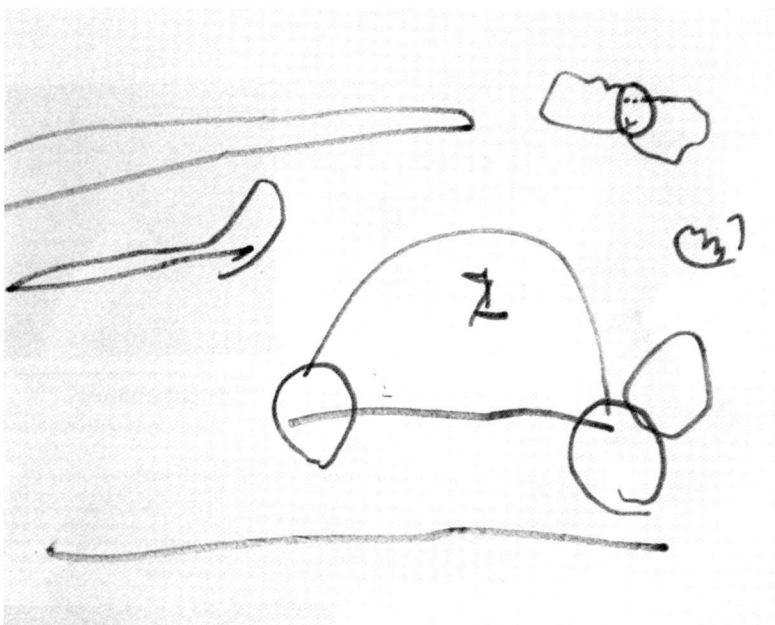

1.34
A small child's repre-
sentation of a car.

The contention here is that people attempt to see complex objects in simple terms to understand their general size and position in space. An object can be fitted into a primitive frame and parts of the object can be handled in the same way at a detail level. Most of what can be seen can be summarised in terms of a few simple geometries, known as 'primitives' (Figure 1.35).

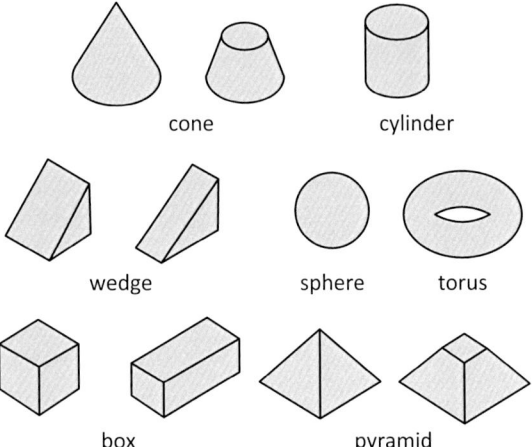

1.35
Many artificial objects can be drawn using these forms ('primitives') as guiding frameworks.

It might very well be that seeing in terms of underlying geometries is an artefact of Western culture. It may even be an artefact of Western art methodologies. However, for our purposes, this is unimportant. What matters is that given this habit of people to see in these terms, a designer may wish to work with this by creating forms that are more easily understood.

The next step in seeing in terms of primitives is that primitives can also be understood as normative structures. That means the object should look approximately like a cylinder or cube. An object might be said to 'want to look like a cube' but it has a part missing or is deformed. If the difference between the object and its primitive reference is ambiguous, the viewer may regard this as a kind of error. The case of unambiguous deviations from primitive forms is discussed in Section 1.5.

EXERCISE 1.7

Find an object in your surroundings such as a chair or electronic device and identify its underlying geometry. Take one element and transform it; for example, make a cylinder longer or wider and alter the other elements to accommodate it. The image shows an electric mixer. Drawing these cylinders from an image requires some drawing skills and analysing how the form is made up. Notice that the handle has a cylindrical shape at the bottom but blends into a wider, noncylindrical form at the top. The main body of the mixer is a cylinder which extends upward to become part of the structure of the upper handle volume. A small cylinder is the basis of the location of the whisk arms (left).

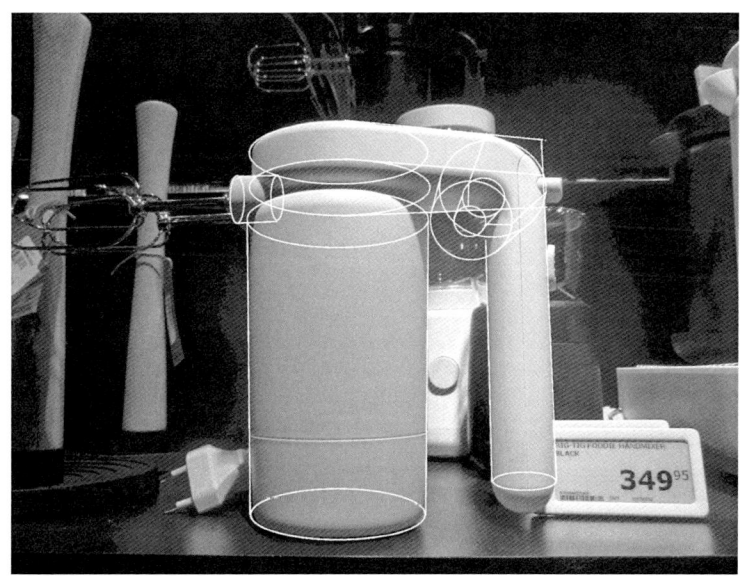

1.5 FORCES AND SEEING 'AS IF'

When you see a streetlamp that's not a straight cylinder, you know from experience that something has happened to it. It's been hit by something – it didn't leave the factory like that. We bring something of this kind of thinking to the way we interpret complex forms. A plywood chair actually has been deformed in a press. A plastic chair was moulded but parts or all of it look *as if* they have been subject to forces bending or twisting them.

This section draws from the work of Cheryl Akner-Koler (1994) on three-dimensional visual analysis and how to analyse forces that seem to have acted on shapes. Akner-Koler's text is readable, handily concise and available online, so readers are directed to look at it in full. In this section I will refer to the part which deals with movements and forces.

Figure 1.36 summarises the ideas of axes, movement and forces.

This discussion of forces rests on the assumption that whether innately or through experience, we learn to interpret the form of objects as if they have been acted on by a force or as if they have undergone a process of change. The object is viewed in relation to basic forms, the primitives from the previous section which have either been transformed by:

* a bending force, or
* cut, or
* being added to.

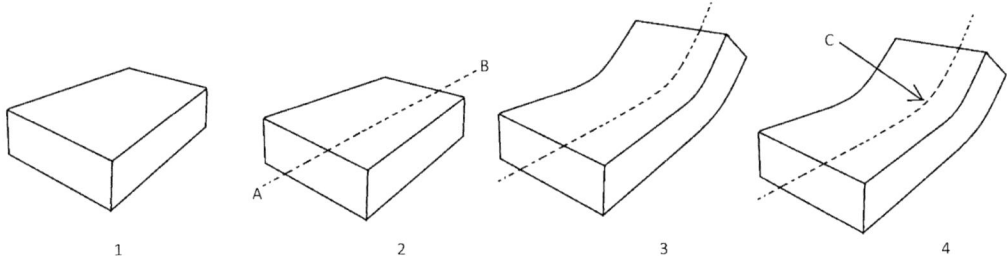

1.36
Picture 1 is a volume such as a box. Picture 2 shows an axis A–B that summarises the box's character. Picture 3 shows an axis with movement (deformation) compared to the box in picture 1. Picture 4 visualises the movement (deformation) as the result of a force C. We see the box in picture 4 as if it has been deformed.

For example, in Figure 1.37, box A may be considered a as deformed version of the ideal box B.

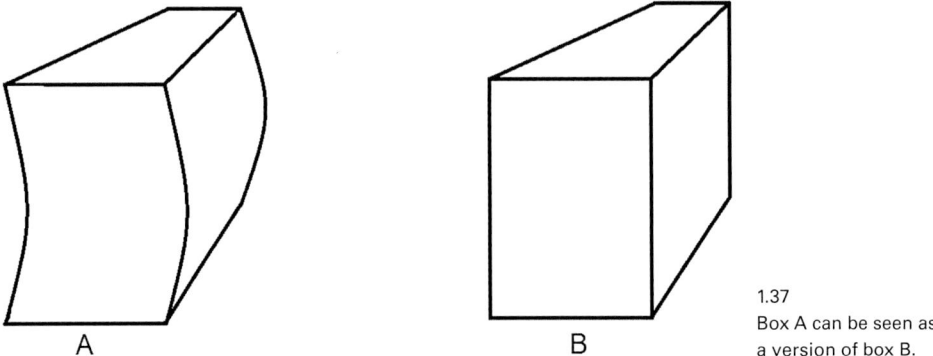

1.37
Box A can be seen as a version of box B.

In Figure 1.38, box 'A' can be viewed as if it has had some part removed, compared to the ideal box to its right, 'B'.

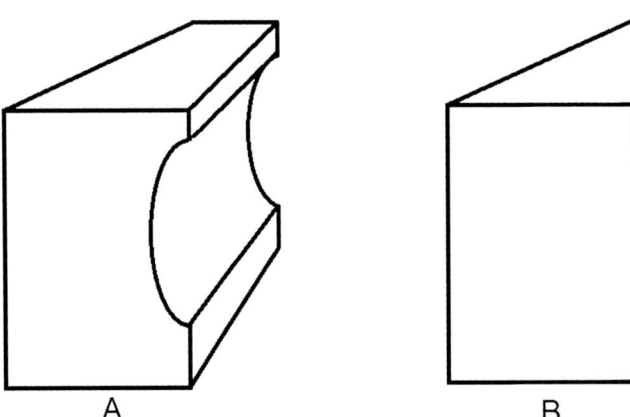

1.38
Box A seems to have been cut, compared to 'ideal' box B.

In Figure 1.38 with the cut box, the effect rests in part on the law of continuity in gestalt theory. We can read the vertical lines on the right side of box A as being interrupted by the elliptical cut. See Figure 1.39.

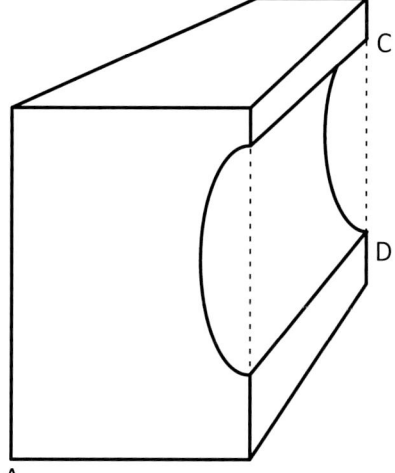

1.39
The vertical lines on the left side of the box seem to be continuous. It looks as if something was cut away.

Akner-Koler began by proposing that we can analyse three-dimensional shapes by thinking of the 'inner- and spatial activity of the elements. These "activities" encompass the combined effects of movements and forces'. Akner-Koler (1994, p. 19) wrote of an 'X-ray vision' that sees the way in which forces have acted on the object. I propose that we can formulate this to say that we see the object *as if* something has happened to it, that a force has been applied to the object to change its form in relation to a hypothetical, simpler version of the form. These forces have a directionality to them. The degree of change is proportional to the amount of force applied. We have an understanding of this from direct experience. A crashed car yields clues as to the violence of the impact. If we see a badly wrecked car, we can infer that the accident was severe. If we see a car with a cracked lamp and detached bumper, we can perhaps assume the impact was not very great. Figure 1.40 shows a mildly deformed block of clay.

To be able to talk about forces we need a spatial structure as a reference. An object can be thought of as having three axes, the X, Y and Z axes familiar to us from geometry. In geometry these are of equal priority. In three-dimensional form we can label them as primary, secondary and tertiary (Figure 1.41). The primary axis is allocated to the longest dimension of the object. The secondary

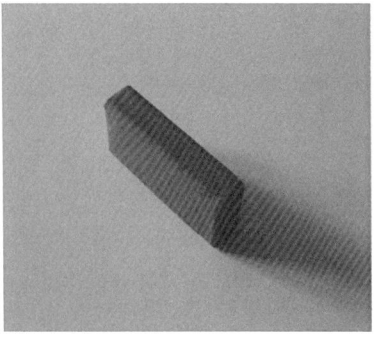

1.40
A block (left) subject to mild deformation (right).

and tertiary axes follow on according to the relative dimension they are paired with. The secondary and tertiary axes may be the same.

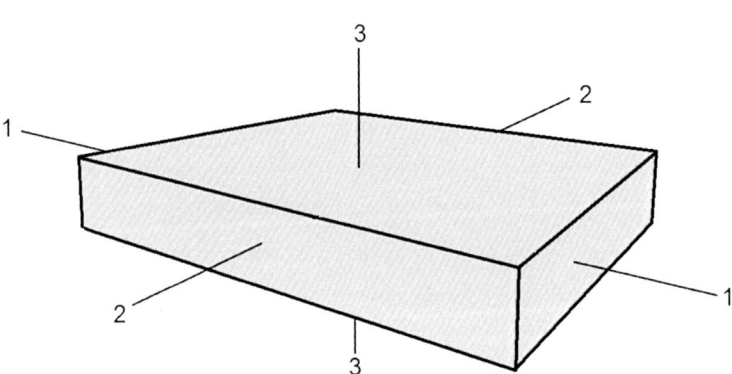

1.41
A simple oblong with its primary (1), secondary (2) and tertiary (3) axes. Is there a gestalt law in operation in this image? Yes. The Law of Continuity – we see the axes as being continuous despite the box in the way. We infer the primary axis is one line, passing through a box.

Akner-Koler (1994) considered forces as if they are acting upon these axes, which creates what is termed 'axial movement'. The example used by Akner-Koler is what appears to be a deformed plane (Figure 1.42). The movement in the axis refers to its path or deviation from a straight line. This 'movement' describes the shape of the surface it is within.

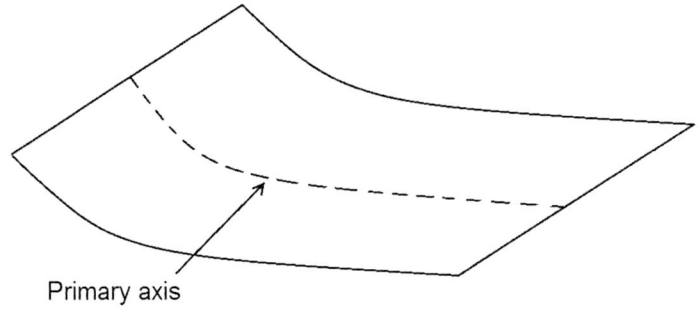

Primary axis

1.42
Deformed plane. Redrawn from Akner-Koler (1994).

This may also be understood in this fashion (see Figure 1.43).

1.43
Another interpreta-
tion of the deforma-
tion: an even force
is applied to the left
edge, moving it up
relative to the right
edge.

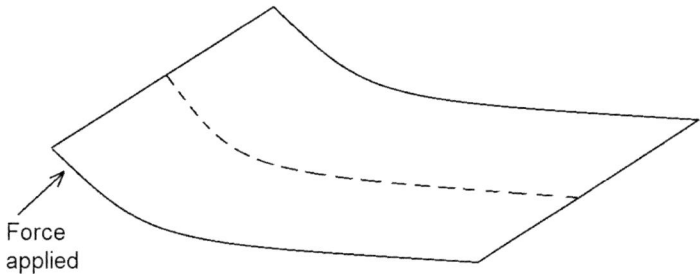

Force
applied

This can be interpreted as though a force was applied in the direction indicated by the arrow in Figure 1.43. In other words, the plane has been bent. The
change is viewed as being relative to the flatter area. That means that we read
the force as having affected the left side but not the right side.

Using the same axial terminology, we can also imagine that an object has
a directionality to it. The great Gothic cathedrals are often explained in terms
of their upward movement. This is because we perceive the primary axes as
being vertical. Frank Lloyd Wright's buildings commonly had a horizontal alignment, being long, low buildings with numerous flat oblongs. They were called
'sideways skyscrapers'. In automotive design (and almost all forms of transport design) the expression is commonly one of implied horizontal movement.
Where the designer wishes to imply speed, this character is exaggerated as
much as is possible within the constrictions of the packages, sometimes even
compromising passenger space to express power and speed (Figure 1.44).

1.44
This Pontiac Trans Am is strongly directional.

Objects with inner axes that are straight might be seen as static or inert (the Gothic cathedrals are interesting because they confound that expectation through their extreme pointiness). The effect of forces is to add visual complexity and, indeed, if we consider again Figure 1.37 with the deformed box A, we see that more information is needed to describe it than the undeformed box. Indeed, it took marginally longer to draw than the undeformed box 'A'. Forces thus give expression and contain more meaning (a surplus of meaning in terms of Nygaard Folkmann, 2013). We can say that the undeformed box is 'just a box', whereas the deformed box can be explained in terms of something happening to it, as if it has a story to tell, albeit a simple one.

Having understood the concept of forces acting on the axes, we can then begin to formulate variations in the forces and also begin to compound the effects. This gives us a vocabulary to discuss the direction, degree and nature of the force. We have dealt with the direction and degree of the force. The nature or condition of the force refers to it being a bending force or curving force.

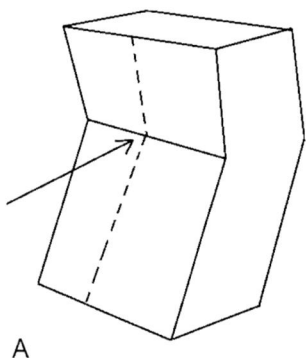
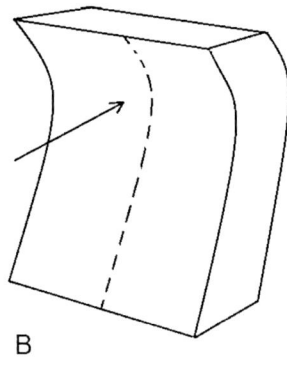

A   B

1.45
Bent (A) and curved axial deformation (B). A is a bending force and B is curving force.

In Figure 1.45, the primary axis of box A is bent and the two faces meet at an angle. The primary axis of box B is curved. These are very simple cases. In more realistic cases the designer will be considering local areas and compounds of forms (see Figure 1.46).

We can be more precise about the nature of the effect of the force. While the direction of the force is simply a matter of vector, the way the axis behaves can be more complex. Figure 1.45 shows the axis bending at an angle or forming a curve, but the axis's curvature can be varied from uniform where it forms a curve of constant radius or it can be some form of an ellipse, or it can be any combination the designer prefers. Figure 1.47 shows a selection of curved axis types and their related forces.

The diagrams are two-dimensional and unambiguous. This is an example using a simple sketch (Figure 1.48).

1.46
The elements of
the chair can be
described in terms of
forces.

Designers are also concerned with three dimensions, which means considering the axial movements in relation to space and, in the case below, the relation to the object's purpose (Figure 1.49).

In the case of this kettle (Figure 1.49), there is a primary axis which shows apparent deformation in response to a force applied (the black arrow). Notice that the curvature increases gently towards the spout. The handle has its own primary axis which is not deformed. The relation of the two axes supports the gentle suggestion that the kettle should be poured. There is a very slight widening where the kettle meets the base part. This can be read as the expression of stability. Here is an example of design for product understanding. There exist quite functional kettles which have neutral axes; that is, they are simple cylinders. The kettle shown has a form which subtly supports the notion that the kettle should

45 ☐

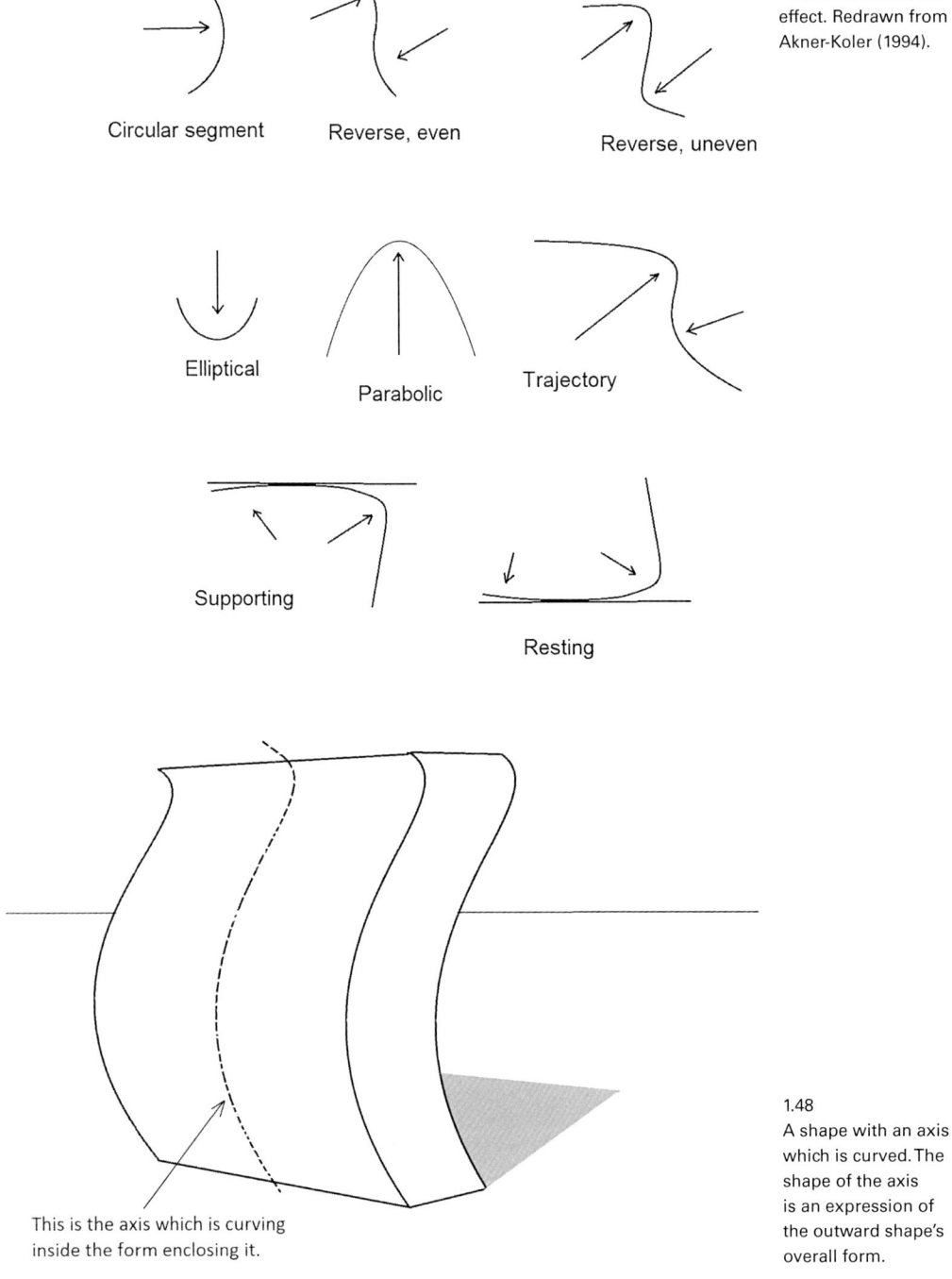

1.47
Forces' direction and effect. Redrawn from Akner-Koler (1994).

Circular segment

Reverse, even

Reverse, uneven

Elliptical

Parabolic

Trajectory

Supporting

Resting

This is the axis which is curving inside the form enclosing it.

1.48
A shape with an axis which is curved. The shape of the axis is an expression of the outward shape's overall form.

1.49
An electric kettle with its subtle accents expressed in terms of axial forces. Which notion from Gibson is in operation here? It's the notion of affordance. The forces are supporting the idea of the kettle's motion upward and of holding the kettle.

be tilted. It also lends the kettle an understated character and is another example of what Nygaard Folkmann (2013) referred to as a surplus of meaning.

These concepts can be used to:

1. analyse the existing designs,
2. make a plan for organising shapes you will make,
3. analyse your work after ideation.

EXERCISE 1.8

The aim of this exercise is to apply ideas from Akner-Koler about the axes and their organisation.

Find an example of a designed object and analyse it in terms of the axes and forces.

1. Mark up the axes of the major volumes.
2. Show the forces acting on the axes.
3. Classify the forces after the schema in Figure 1.47.
4. What is the relation of the axes to one another and to the directions up, down, left, right, etc.?
5. What do you think the intention was to organise the axes as they were? Is the form dynamic or static? Is it directional?

Remember that graphic elements may be superimposed on three-dimensional forms. The boundaries of these may also be analysed in terms of forces. 'Graphic elements' means boundaries between one part and another, especially parts with different colours.

SUMMARY

This section has presented several ways in which objects can be viewed. Gestalt theory provides a set of 'laws' about interpreting lines and groups of objects. It makes overall sense when understood as the rules for establishing the relation of objects to their environment. The theories of Gibson and Gregory establish how information is processed by the brain's visual systems. Gibson makes us aware of the affordances of objects; that is, which actions the shapes permit or exclude. Gregory alerts us to the idea that a design should support rapid interpretation. Anthropomorphism explains the tendency to read inanimate objects in terms of human forms and behaviour. The section on primitive forms is an argument that the mind's capacity to understand three-dimensional objects depends on their being analysed in terms of simple geometrical forms and compounds of these. Seeing 'as if' depends to some extent on transferring our experience and understanding of materials and forces to interpret objects of fixed form. The axes indicate the underlying structure of a form or part of one.

Being aware of these concepts allows the designer to design for product understanding through the management of forms so that their meaning and function is more likely to be understood by the user in the way the designer wishes. This allows us also to see that while design and art have a concern for the visual, the designer is interested in controlling forms to reduce ambiguity, whereas an artist might actively want to maximise it.

NOTES

1 For detailed insights on that, see Bruce et al. (2003) or Ware (2012).
2 Note how the boxes are connected with directional arrows; the arrows are explained with a short phrase which makes a statement of the box-to-box link. Not all possible connections are shown.
3 The subject matter is complex, and I hope specialists in visual cognition will not be offended with the license I have taken to simplify and compress the material for this chapter.
4 It doesn't look as if it's meant to be pulled. The shape does not afford pulling but perhaps pressing or rotating. If you need to add a label, the design has probably failed.

FURTHER READING/KEY TEXTS

Akner-Koler, C. (1994) *Three-Dimensional Visual Analysis*. Stockholm: University College for Arts, Crafts and Design. **I can't recommend this important text highly enough. It is essential reading for designers**.

Confer, J.C., Easton, J.A., Fleischman, D.S., Goetz, C.D., Lewis, D.M.G., Perilloux, C., & Buss, D.M. (2010) Evolutionary psychology: Controversies, questions, prospects, and limitations. *American Psychologist*, 65(2), 110–126.

Epley, N., Waytz, A., & Cacioppo, J.T. (2007) On seeing human: A three-factor theory of anthropomorphism. *Psychological Review*, 114, 864–886.

Gibson, J.J. (1972) A theory of direct visual perception. In Royce, J., & Rozenboom, W. (Eds.). *The Psychology of Knowing* (pp. 215–227). New York: Gordon & Breach.

Gregory, R. (1974) *Concepts and Mechanisms of Perception.* London: Duckworth.

Norman, D.A. (1988) *The Psychology of Everyday Things.* New York: Basic Books. **Another key work for designers**.

2 Resolving the constraints

2.0 INTRODUCTION

Every design is a failure, wrote David Pye[1] in 1978. By that he meant that no designed object can do everything, satisfy everyone, last forever or cost nearly nothing. Military projects have a very low tolerance for failure but they cost huge amounts of money. Even then, the laws of physics limit what the military can design. At the other end of the scale, a disposable plastic cup costs little, holds water and is probably a horrible item. Most things lie somewhere between astronomical cost and next to zero cost and reach a compromise by failing in an acceptable way (for some users – not all of them).

This chapter discusses the idea of design as involving unavoidable compromises between the primary needs of function, materials requirements and production methods. One of the three is usually a primary driver affecting the design result, or appearance. Choices about appearance equally affect the choices about function, material choice and production methods. No design can avoid this balancing of constraints. What is called 'uncompromising' design is always a compromise, whether in cost, appearance or functionality.

Military applications are a special case where the compromise is the cost – usually immense. Even these projects get limited by time (and political patience). This chapter also discusses the vexed philosophy of functionalism which accords respect to the functionalist ethos but also draws attention to its limitations and asks the reader to consider a wider interpretation of what functionalism might mean. The chapter presents a discussion of how initial choices about style affect the resultant object in terms of material and manufacturing methods. The examples of limited production and mass production make clear the two poles of cost versus quality.

The main points of this chapter are illustrated in Figure 2.1.

As stated in Chapter 1, the form must be understandable. If the other factors such a price and material compromise form, the designer should try to either a) make clear to the customer the reason the form is compromised or, better, b) argue more strongly with colleagues for a change in parameters to preserve the product's appeal. As I tell students during classes on this topic, the customer doesn't know about or care about many of the demands faced by

DOI: 10.4324/9781003183303-2

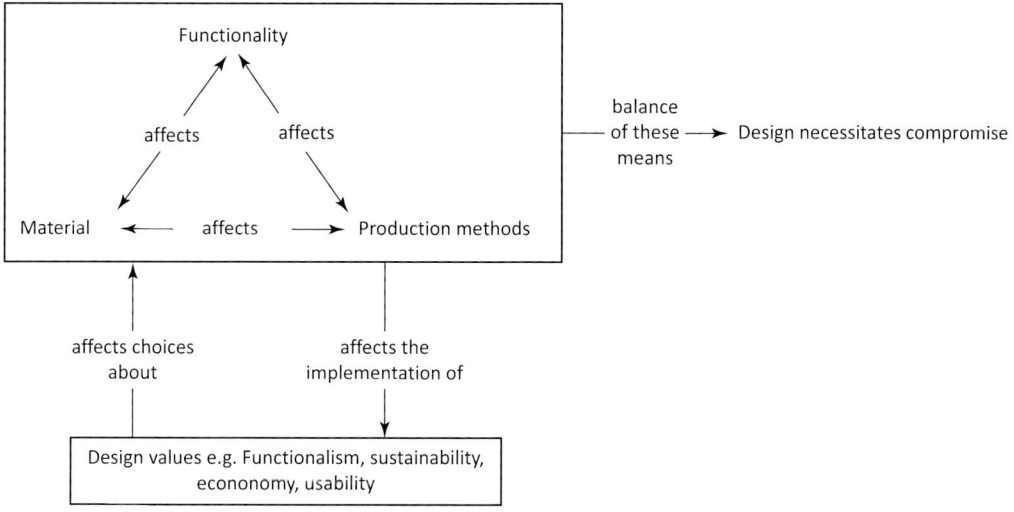

2.1
How the elements of this chapter are related.

a designer in solving a problem. They can only see the product in front of them and it ought to look as good as possible.

In Figure 2.2 there are three different curves. It is not sufficient only to be able to identify which one is of the highest quality or most pleasing. Often the choice is between several forms of equal mathematical geometric validity and visual appeal. A designer needs to have a rationale for choosing which curve or surface makes most sense. This chapter presents some of the underlying arguments concerning the factors affecting the form of designed object.

2.1 DESIGN NECESSITATES COMPROMISE

This book focuses most on form and line, the aesthetic dimension. It is, however, not divorced from the other constraints of design. The matter of cost is often left unspoken when discussing industrial design. David Pye (1978, p. 70) wrote, 'Of the many inevitable conflicts between the requirements of economy the crudest is that between durability and low first cost'. Every design has some element of cost requirement built into it, even design which falls under the category of 'art design', such as conceptual projects by critical designers Dunne and Raby.[2] The item must be affordable, even if it only has to be made once. Budgets are ultimately finite. At the lowest level, the object can't be so cheap as to be ineffective, though there are products which barely reach the level of basic utility. Even finding out the basic level of utility is costly, and it is often cheaper to approximate it, Pye noted (1978).

Assuming basic expectations of utility are met, one presumes then a reasonable cost relative to the parameters of functionality, appearance and quality.

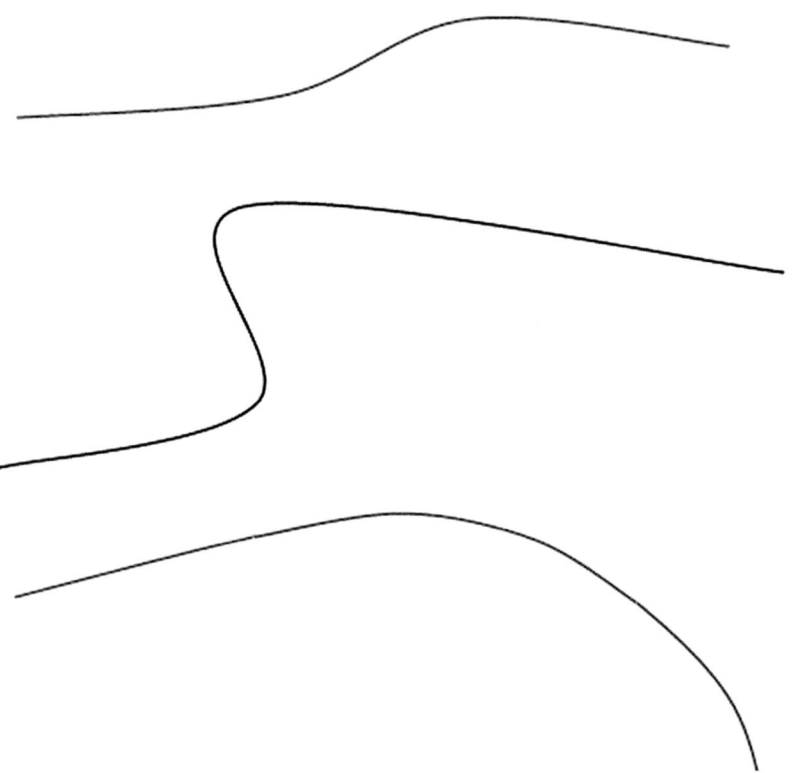

2.2
There are two fairly decent curves here (and one not so good). Deciding which one to use means more than only being able to assess curvature quality in isolation.

Quality here refers to material quality such as whether it is made of the right material and assembled neatly. These interrelated considerations are of direct relevance to form because

1. when the constraints are properly considered, the chance of achieving a good form is increased.
2. these factors ultimately determine the meaning of the form.

So, while this book is chiefly about form, it is necessary to underline here that form does not exist in a vacuum. Figure 2.3 shows four parameters driving design. Other parameters could be added but at the risk of redundancy. For example, sustainability concerns could be added. However, they can be seen as an element of functionality, depending on how one defines that term. It could be that one function of the object is to be energy efficient. Or a function of the object could be that it is to be easily recycled. *Function* is a broad term and here means what we wish the object to do over and above its direct job such as cutting or cleaning. The term *material* can be subsumed under *appearance* given that forms have to be made of something. Notice that the four factors aren't arranged hierarchically. Quality might very well include appearance if quality includes the degree to which the shape is pleasing as well referring to the standard of material and

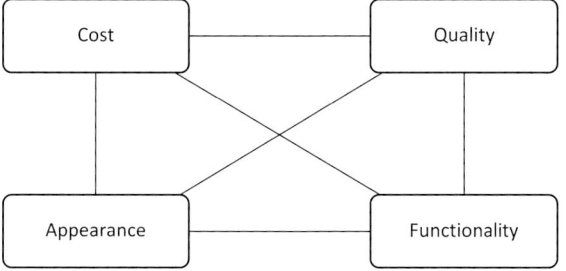

2.3
Everything affects everything else: cost, quality, appearance, functionality. Material is not explicitly included. It is here subsumed under appearance (for the right look) and quality (for the right durability). What means exists to assign priority to these demands? A design process is how one finds a path through the conflicting demands of a project.

assembly. These diagrams are not meant to be definitive but illustrative of some of the considerations required. A definitive diagram could only be made at the end of a design project when all the work has been completed.

EXERCISE 2.1

This exercise is to help you consider the impact of different priorities on the outcome of a design process.

Find designed objects and sort them into four categories according to whether you think

1. cost,
2. quality,
3. appearance, or
4. functionality is the strongest factor driving design.

Are there cases where cost *should* be high and not low?

You could then do an analysis of how you could change the emphasis of the product to see the design implications. For example, if a piece of furniture is made for low cost, what aspects of the object would change if you were to bring it to a higher price point and make it visibly more appealing? Covering it in gold leaf would not be a good solution as it would add cost and not much functionality. Look at the joints of the furniture: are they simple? Are they flush? Do they interfere with or confuse the overall appearance? What is the material? If it is a flat-packed bookcase the material may be veneered fibreboard.

Using sketches and sections, redesign the joints so that the object is not flat-packable but sold fully assembled. As a counterexample, find a designed object and identify how its appearance would be altered to make it much cheaper to manufacture.

You could use this four-way schema to analyse products in a market sector you are designing for.

Architects face similar constraints to industrial designers but with different emphasis. Materials in construction and appearance are very strongly correlated in contemporary design. Using an example of a building, analyse the interrelationship of the form and the materials. If it is made with cast concrete, what would the effect of using wood or brick be? Where does functionality come into the equation?

Industrial designed objects normally reveal their function through their form. Buildings, in very simple terms, have their functions enclosed in a waterproof 'skin' which may have no structural role. Some buildings do reveal their function – how? Is it symbolism? Is it structural?

The appearances of internal spaces have a loose correlation to function: a room can be an office, store or a sitting room or all three. Is the relation of form to function different in architecture and in design?

Graphic designers tend to be less concerned about cost and material quality, yet they are also concerned with appearance. How is graphic design compromised?

Urban planners plan spaces on a very large scale. Does 'appearance' seem like a priority for urban planners? Perhaps it does matter in city centres. How about the layouts of our suburbs? Take a look at a suburban environment and consider the priorities that are apparent. Our urban spaces are also the result of compromise. Can you see where these have been made?

Though the four parameters in Figure 2.3 stand separately and are supposed to represent open, independent variables, this could be a little misleading.

Taking the parameter 'appearance' as a starting point, tied up with 'appearance' is a set of implicit constraints.[3] It is not a freely variable parameter. Since industrial design has been around for more than a hundred years, it has developed a form language that is verging on the traditional. This includes rounded edges, design for assembly and part split lines that assume, say, construction from injection moulded plastic parts or pressed steel. So, to start with an idea of a preferred shape (as one often does) means automatically ruling out the likelihood of certain materials and processes.

The same is true for architecture and fashion. There are certain conventions in these fields which strongly affect the influence of the other constraints. In architecture these conventions might include

1. the assumption of straight edges (easier to make: consider Figure 2.4),
2. a reduced number of parts (simpler to make: consider Figure 2.5),
3. the use of conventional materials, which in turn relates to (1).

2.4
The design of this wooden-cased 1960s radio is strongly related to cutting processes. (Image: author's collection)

2.5
This Apple plug is made from injection moulded plastic. Its aesthetic is related to a cheaper and simpler production process, but note the high quality of the curvature of the corners and edges.

2.6
Tour Vercors,
Grenoble (1966). In
this case architects
have decided to
show the living
modules as distinct
units. The wish to for
multi-story construc-
tion determines the
material. It could not
have been made of
brick, wood or stone
and still looked like
this.

Taking a simple product as a case, let's look at the implications of prioritis-
ing the constraints in various possible ways. In industrial design, to draw an
oblong shape with rounded edges strongly suggests something made from
plastic and so the decision about the preferred shape has almost entirely
ruled out carved wood or stone. You could make such a thing out of wood,
but the cost of this item would probably be prohibitive. Taking material as
a starting point, and assuming wood is to be used, then the form and cost
factors are drawn into play: the object will have a simple shape and probably be costly. With functionality as the start point – for example, this object
must be ergonomically efficient – the appearance has to a quite large degree

been determined and many materials ruled out. The point is that while one can't know in advance the *precise* interplay of these factors, when a factor is picked to be a dominant factor, you will quickly design around some general assumptions that follow closely.

Another misleading character of Figure 2.3 is that it does not include a box for feasibility, which is about whether the design can be economically produced in series: the 'designer decides on the form which is realised as a product *via appropriate manufacturing processes*' (Karana and Hekkert, 2010, emphasis added).

It should be clear now that the problem is that one could eventually add more and more boxes to the diagram to reflect the nature of design as a synthetic discipline. However, I wish here to focus on these four parameters. In this instance (Figure 2.3), the boxes' 'cost' and 'functionality' may be said to account for feasibility if one assumes a) the production method must be affordable and b) that functionality includes the function of being makeable as well as being usable.

EXERCISE 2.2

The aim of this task is to link constraints to an existing design or, better, one you have done yourself.

Write down all the factors that might affect the appearance of a design that you can think of. Then try to assign them to the four boxes in the Figure 2.3. For example, one factor might be 'the colour preference of the users'. That would go into the box about form. A material choice could be put into more than one box since it affects form, functionality and quality. It might be just a matter of a final surface finish, though. Think in relation to a design project you have done or are doing.

The learning objective of this exercise is to see the underlying structure of the many factors that affect design outcomes. Are there some factors that don't fit into the categories? Are there some factors that seem so important they should not be hidden among broader categories?

Is your project cost driven? Is it driven by function? Or material? Or by appearance?

The interesting and, indeed, slippery thing with design is the business of balancing these (and many other) constraints. It is about finding out how to reconcile them such that the desired appearances don't demand too high a price or that the price requirement does not result in aesthetic or other deficiencies. For the customer, the appearance/price compromise is asymmetric since the price is represented by an easy-to-understand number on the price tag, whereas the appearances are subjective and may take some time to understand and appreciate. Successful design is often that which overcomes consumer resistance based on price. Successful design is also that which increases the chances of the customer liking and retaining the product for its entire service life (or beyond).

Taken as a whole, Figure 2.3 encapsulates the four-part nature of industrial design: 1) low unit costs and 2) styling driven by materials that are 3) easy to work with in 4) large quantities.

And, further, the product has to look comprehensible and attractive to the user. Graphic designers are less concerned with low unit costs and perhaps more about how much time and money the client has offered for the job. Architects may be concerned with low unit costs at the level of building element: windows can be made to measure or ordered from a catalogue. At a large scale, the cost factor can be understood as being about how many of the features can delivered at the right level of quality given the budget. There is always an element of compromise in solving the problems of design tasks. Those working with fashion thus seem to have more in common with product designers than with architects and graphic designers. There is a wider gap between form and function, which is why there is so much expressive potential in fashion.

EXERCISE 2.3

The aim of this task is to show how first choices about form constrain many other aspects of a design.

Form is what the user perceives in the first instance when confronted by a design. After the user has had some time to understand the product, the feelings and impression of form are what is left. For product designers, form is often the most dominant factor in their considerations. What happens when you make conscious efforts to prioritise form?

Try to find designs which are predominantly form-led and investigate the effects on performance and other attributes. You could take a project you are working on and spend some time reworking it so that it is as attractive as you can make it – what, if any are the effects on the other factors?

Note: form is often considered in relation to other forms. One modern chair amidst a room full of old-style wooden chairs might appear striking. A room filled with 150 identical modern chairs might look banal. You might wish to set your design example in the context of others of its type to help assess the effect of a strong emphasis on pure form.

Learning outcome: through this exercise you may be able to see creative potential in drawing unconstrained by practical considerations. You
might also better understand the consequence of this approach. You
might also see how much of your design choice is driven by reaction to
existing forms rather than emerging from the design problem in isolation. Marketing often demands designers invent innovative forms that
compromise function or really aren't as satisfying as earlier attempts to
solve the same problem.

2.2 FUNCTIONALITY: HOW THE OBJECTS
WORKS AND IS SEEN TO WORK

A major ideological battleground in design relates to the primacy or otherwise
of functionality. There are a number of approaches to this. The most important,
practically speaking, is functionality understood in ergonomic terms and how
well the device does its job. Ergonomics deals with the cognitive and physical fit of the object to the user. This is a comparatively straightforward matter
inasmuch as it is dealt with by a well-developed body of literature (Dreyfus,
1967; Bridger, 1995; and numerous academic text such as Lewis and Narayan,
1993; Broberg, 1997; Kanis, 1998; Han and Hong, 2003; Liu, 2003a, 2003b;
and many others). An object could be ergonomically excellent but not look as if
it is ergonomically excellent (or ergonomically awful either).

At the next level is design to communicate functionality (see Figure 2.7).
The previous chapter presented a general introduction to the work of Gibson
and the concept of affordances which indicate what the object can do; for
example, a spatula affords scraping and spreading. Ideally, a product should
be shaped in such as a way as to make transparent what it is for and how it
works. A chair has a horizontal surface that affords sitting and a door handle has
the form and position to afford being grabbed. Further clues in the door handle design indicate whether the handle should be twisted or moved in some
other way. These are very simple cases in which the object has one function.
However, many objects are a compound of forms and may have more than one
function. A control panel would be a more complex example.

The urban environment or a building's interior can also be understood
in terms of affordances. The image in Figure 2.8 shows a street where the
options available to the pedestrian or driver are quite clear. This is because the
vertical surfaces of the buildings make it obvious how the roads are laid out.

More complex still is the ideological tension between form and functionality where the appearance or style of functionality is at stake. For more on this,
see Section 2.5 on the functionalist ethic. The matter of functionality and form
moves into the category of appearances because the topic now falls under
the category of functional*ism*. That is, that a designer styles the object so it

2.7
The remote control (left) makes no obvious concessions to ergonomics. The visual order is the sole aid to the user. The telephone (centre) has quite large buttons with clear labelling so the basic functions are clear. The device is also slightly narrower at the end to afford holding it (it is narrower at the top end to look symmetrical). The mouse (right) is subtly ergonomic – it is smooth to fit in the palm and has a thumb rest.

2.8
The close relation of the building footprint to the carriageways means this street has visible affordances as to the directions pedestrians and drivers can go. Some of them are marked up with arrows. Are there any affordances that could be added?

communicates its functions while also making clear the conscious effort taken to avoid decoration or over-expression. To understand this difference, take a look at the two chairs in Figures 2.9 and 2.10. Both are equally able to provide place to rest for the user. Figure 2.9, the design classic Wassily chair, is minimally decorated, whereas the other (Figure 2.10) is far more decorative.

2.9
Wassily chair by
Marcel Breuer.
(Courtesy of Knoll,
Inc.)

There is no real sense in which the Wassily chair is actually more functional than the decorative one. It is not less functional, either. But the Wassily chair is designed to makes its function (seating) a core element of its style. The other chair has the function of conveying the status of its owner through the use of expensive, ornate craftsmanship as well as being a seat. The chairs reveal the distinction between styling and decoration. Both chairs are very thoroughly styled as in they have been subject to intensive consideration regarding their appearance. One chair is much less decorative than the other.

2.10
Decorative chair,
c. 1920.

2.3 MATERIAL: THE IMPACT OF MATERIAL IN RELATION TO FORM

Material is in some ways like colour. Every object will have a material and a colour, yet the exact material and exact colour are things designers often tend to leave 'until later'. Material can be considered as either of first importance in design or seemingly something apparently incidental. While discussing the importance of material, researcher Anders Haug (2018) wrote, the 'associations or meanings are dependent on the context in which the material is used'. Plastic is an entirely appropriate material for a bucket and for much of an aeroplane interior. A chair made entirely of plastic is risking being seen as very cheap and of low value. It is largely a matter of relative values and not absolute values. Writers Hekkert and Karana (2014) even went so far as to say that materials do not have a meaning at all. They then qualified this by noting that, for example, steel and stone are viewed as cold but also high quality; they wrote that roughness is associated with being natural, and soft materials are viewed as alive, in contrast with dead, hard materials.

Material, like colour, can be entirely abstracted from the essential geometry of a designed object. However, in the end, a material must be chosen: 'Product material choice affects how a product will be manufactured, how it will function and how it will be experienced' said Haug (2018). How it will be manufactured will affect the forms possible, from the small scale of the exact curvature allowed by the tool's draft angle[4] to the gross form (highly sculpted versus minimal geometrical shapes).

When material is a matter of secondary importance, one prepares a sketch of a rough idea of the object. Then, through consultation with production engineers and by means of prototyping, one sees how close one can get the feasible version to resemble the design intent (Van Bezooyen, 2014). This is what is called a design-led approach: 'going from macro-requirement to a specific material with a particular microstructure' (Ashby and Johnson, 2007). Put another way, the choice of material will have consequences for the appearance of the object at the finest scale.

This treatment of material is possible because the usual forms for industrial design (rounded edges, draftable shapes, cuttable shapes) have been arrived at after decades of use and are often intuitively understood. What was once the result of careful experimentation is now effectively a traditional style. See Figure 2.11.

So, although designers can sketch forms without knowing exactly the material to be deployed, the kinds of forms used in industrial design suggest known materials. As Karana and Hekkert (2010) noted, people associate particular materials with particular forms. The reverse is also the case: 'These associations are mostly due to the prevailing use of a material in a certain shape used in daily experienced products'.

However, even when material is of primary importance, its effect on form is still somewhat weak. This is because designers have a reasonably reliable

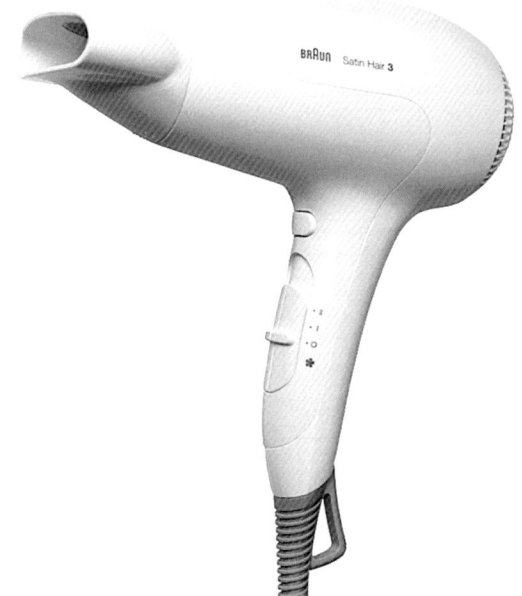

2.11
A contemporary
hairdryer. (Courtesy
of Braun)

understanding of which materials to use for which task and many forms can be achieved using a variety of different processes. It is only when cost is taken into account that the form/material relationship becomes stricter. It is very seldom that the choice is between leather or glass, which would indeed result in two different forms for a lampshade, for example. A more familiar kind of design choice is perhaps between plastic or metal or between plastic and glass or which metal to use out of a preferred range. To choose a noble material is very often a conscious decision made at the start of the design process in the light of the client demands for high-end products.

Drawing on the ways of seeing outlined in Chapter 1, the choice of material affects the way in which the form is understood. The section on forces discussed how a form can be understood to be the result of forces that appear to have acted on the object (or parts of the object). The same sort of thinking applies literally to material in that it has been shaped by an industrial process to achieve the desired geometry. Plastic is typically moulded. Metal can be pressed, cast and cut. Wood can be cut from whole or may be processed into pressed or moulded forms. Typically, the form of the material then is in accord with the way it has been processed.

EXERCISE 2.4

Referring to Table 2.1, draw a random shape or a few random shapes. Then find out which material and which process would be most suitable to make it with. In the example shape below, it looks as if it has regular profiles at the front and back (A and B) at right angles to a flat face, labelled C. It could be made of wood, metal or plastic. Wood seems most likely as this kind of shape seems too simple to demand making a mould for plastic or a die for metal. You might want to ask a tutor for assistance with this exercise or to at least have a look at a book on materials and process. *Industrial Design Techniques and Materials* by Raymond Guidot is a good one.

Table 2.1 Some materials, their typical processes and typical resultant forms.

Material	Processes possible	
Plastic	No strong form restrictions but the form must be mouldable	Drafted 3D shapes
Wood	Carved, cut, milled, turned	Straight edges, profiles, sculpted
Metal	Cast, pressed, cut, milled, turned, blown, extruded	Drafted 3D shapes, regular edges, profiles
Glass	Blown, poured, spun, moulded, cut, extruded	Blown and spun 3D shapes, profiles, planes
Fabrics	Cut	Tailored assemblies, planes, cylindrical forms

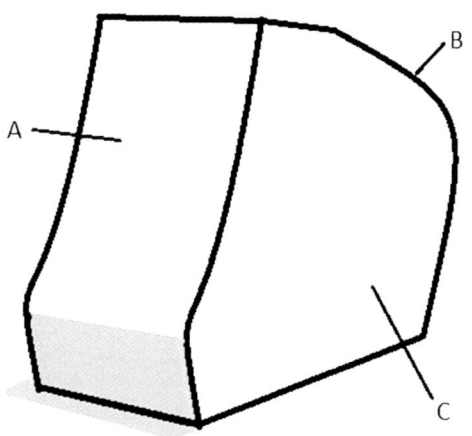

Where a design may run into problems is when the form-giving and material choice are not in accord. This notion of accord rests chiefly on the notion of truth to materials and depends on judgements that are not reducible to simple rules. An example is the use of surface coatings to make one material resemble another. An obvious example is the use of metallic coatings which make plastic look like metal or alloy (see Figure 2.12).

2.12
A plastic part coated to look like alloy.

If we take a strict view of truth to materials, this example constitutes a case of design dishonesty. The case against this dishonesty is that the material is not what it purports to be. The user is being promised something that that object can't fulfil. Spelled out, the promises is perhaps 'I am made of metal and will behave like metal; I am expensive'. The promise can't be fulfilled and will be seen to be so when the metal-effect coating wears off or when the plastic cracks when subject to a force metal would resist.

A similar argument can be made about the use of wood-effect finishes. These used to be more common up to the late 1990s. These days, imitation carbon fibre finishes are a more likely feature. I will discuss the case of fake wood finishes. The case against the wood-effect plastic is, first, that it is not what it purports to be but also that the shape itself may be implausible. Wood in car interiors was originally used in the form of veneers which were flat or gently curved. Some costly vehicles used wood pieces that had been carved.

A wood-effect component that looks as if complex hand-finishing would be needed to achieve that shape is, like the metal-effect finish, making promises the item cannot live up to. This reading of the wood-effect component depends on an intuitive understanding of wood's untreated form and what might be required to transform it into a delicate-looking piece of interior trim. That intuitive understanding is similar to the one discussed in Chapter 1 where the viewer interprets the form in terms of what might have happened to it to give it the shape it has. In Chapter 1 the transformation was explained in terms of forces. It could also be in terms of the cuts required to trim the excess material.

The designer's response to this kind of problem is to close the explanatory gap by either changing the form to make it plausible for the suggested material or selecting a material finish appropriate to the material used. Both of these responses have their own problems. If, for example, the form is simplified, it may not fit in with the forms of the objects around it. Or, if the material is not intrinsically appealing, the overall effect of monochrome materials may be unsatisfactory, too. Some users would prefer some textural and colour variation within a larger context of monochrome components. Other users value visual simplicity. Finally, some users are indifferent to the finish. For the first group of users, wood-effect or metal-effect finishes are not at all troubling because they are not concerned with truth to materials. For the second group, visual uniformity is the price to be paid for honest use of materials. Where a variety of finishes can be offered, all groups can be satisfied. Where there is only one product, the designers will be forced to determine which group is larger and aim to satisfy them.

Some research has been done on the relation of material to form and the results are ambiguous. Karana and Hekkert (2010) investigated the relations between the user, the material and the product using plastic and metal objects as their test cases. Their work indicated that 'the type of product and the way the material is shaped have a big impact on what the material expresses'. Karana and Hekkert also found that 'females were more sensitive to variation in materials than men. In other words, whether a product was made of metal or plastic made a greater difference to the evaluations of females'.

More interesting, and confounding, was their finding that the 'meaning of a material can change in different products; it can be different for different people of different cultures, in different contexts, at different times' (Karana and Hekkert, 2010). They recommended careful context-dependent analysis of the meaning of a material rather than the design 'making material decisions based on gut feelings'.

The general conclusion one can draw from considering material is that, other things being equal,

1. the material choice is somewhat arbitrary once one considers the menu of options existing in industrial design, so that
2. other factors tend to drive material choice.

As a general principle, preferred shape (form-led selection) tends to dominate material choice. The shape then has moderating effect on the meaning of the material; for example, aluminium connotes cheapness when used for drinks cans but is an exclusive material in automotive applications (though it is usually coated). Wood used in packing crates is seen as cheap but wood used for, say, a radio casing, would be perceived as high value. You can see how decisions about material and form can be related in the flow chart in Figure 2.13.

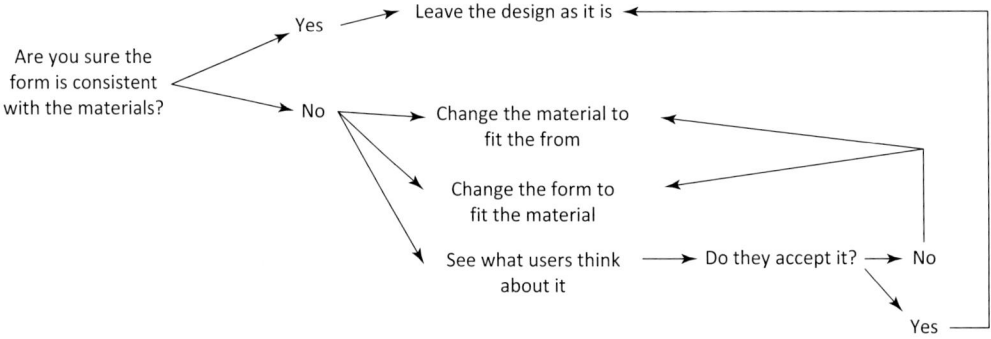

2.13
A material and form decision-making flow chart. For it to work you need to know what shapes are consistent with the material in question. You could also ask users at the very start whether they are happy with the material and the form. Many users probably won't mind very much. This kind of issue is of more concern to professionals and critics!

Looking specifically at form, the biggest factor affecting the interpretation of material is in terms of its plausibility and its use as signifier. A low-cost hi-fi device with fillets of alloy-effect plastic can create a disturbance in the mind of some consumers. They might find it unlikely that alloy would be used for a low-cost appliance. However, other customers might find the visual reference to actual alloy pleasing or might simply like the appearance of alloy for its own sake. Not everyone is very fastidious about truth to materials. Grained, dark plastic might serve the requirement of functionality and durability for a car interior, yet some users might regard it as too utilitarian. As with many of these questions, the judgement and experience of the designer may need to be supplemented with user inquiries.

In relation to sign-posting the object's function, material choice can be an effective means to direct the user to certain aspects of the object. A smooth plastic button may work as well as a matte, rubberised one, but the latter has the advantage of visually emphasising the role of the part. If it is not possible to use a special material to emphasise function, then more work must be done by the form to convey the message of the part; that is, that it should be gripped, touched or pressed. An example of this is the brightly coloured handles in public transport. They are not that attractive, but through the use of colour, the user can see more easily what they can hold on to.

EXERCISE 2.5

The aim of this task is to examine the relation of form and material.

Using one of the objects selected in Exercise 2.1, analyse the role material plays in the form of the object. What would the object look like if a markedly different yet appropriate material was used? An example might be a mug made of steel instead of ceramic or a lampshade made of paper and not plastic. What effect would this have on the gross form of the object?

Conversely, choose an example of a sculptural artwork and determine what changes would be needed to prepare it for mass production from a commonly used material at a 1:1 scale; for example, injection moulding (consider draftability) or steel pressing (draftability matters here, too).

Alternatively, choose a dining room chair and redesign it for another set of materials, assuming a reasonable cost. A simple rule for this exercise is that no part in the redesigned version should be made of the same material as the original. If the legs are metal in the original, don't use metal.

2.4 FORM AFFECTS PRODUCTION METHOD AND PRODUCTION METHOD AFFECTS FORM

As stated in the introduction to this chapter, the desired style constrains the material and manufacturing methods. Consider, for example, the appearance of a Georgian candlestick and the way it was made (see Figure 2.14, left). Then consider the appearance of a modern vacuum cleaner (see Figure 2.14, right). While in principle both could be either mass produced or handmade, the ever-present vultures of cost and quality fly overhead. A mass-produced Georgian candlestick would be a very much inferior proposition (when seen up close). The look makes promises the details won't fulfil; for example, visible tool split lines and a finish which will wear off eventually. A handmade vacuum cleaner may not even work but would certainly look strange if carved from some kind of wood. Perhaps the body could be made of pressed metal, though the panel gaps would doubtless be larger. The cost would certainly be much higher than the plastic version.

The examples of handcraft production and mass production make more evident the two poles of cost versus quality. Alert readers will observe that cost and quality are not always directly opposed: the remarkable thing about mass production is that it makes possible both low cost and very high quality. Most domestic appliances are affordable and built to a standard that often far exceeds adequacy. Modern clothing is much more affordable than that of 1900, is more comfortable to wear and performs better. Two aspects need to be remembered here, though. One is equivalence. Most domestic appliances were not technologically feasible in the era of mass production. The equivalent of a washing machine in 1850 was a person with a large wooden tub and a bar of soap. The second factor is relative quality. It would be wrong to compare a standard modern person's clothing to the clothing of an aristocratic person from 1850. But if we compare haute couture with aristocratic clothing from 1850, the 1850 clothing would still compare quite favourably. If we consider the clothing of ordinary citizens then and now, the 21st-century citizen comes out ahead by having cheaper, lighter and more comfortable apparel.

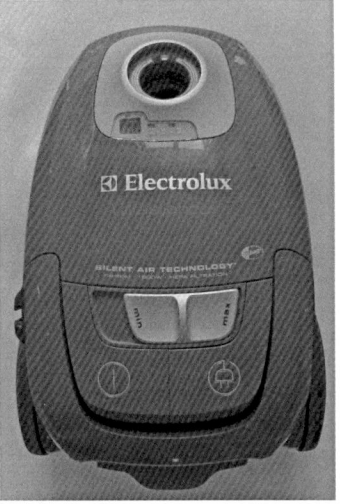

2.14

Georgian handmade tableware (left) and a mass-produced vacuum cleaner (right). (Left image: Credit: By Sean Pathasema/Birmingham Museum of Art, CC BY 3.0, https://commons.wikimedia.org/w/index.php?curid=18805475)

EXERCISE 2.6

This exercise is about the effect of mass production methods on form-giving.

Consider the implications of a) making the vacuum cleaner in Figure 2.14 by hand and b) of mass producing the silver tableware. What effect does it have on the form if the production method changes? Notice that making a mass-produced silver teapot might be less of a challenge than making a vacuum-cleaner body by hand.

In architecture, mass production applies to some elements such as bricks, window elements and structural members like rolled steel joists. It is in the assembly that hand production becomes necessary. Have a look into attempts at mass production of buildings – there are a good number of examples. An early one is the Pacific Ready-Cut homes of 1908–1942 which were built in California. What effect do you think mass production has on the forms? Another way for architects to make building lower in cost is system construction where large subunits are made in advance and assembled on-site. The architecture of the 1960s features system-built elements and was applied to high-rise buildings. The appearance was linked to the style of the era with structural elements visibly expressed. What are the consequences of this?

Another subtle point emerges when we consider form and production. The vacuum cleaner is as much a product of a design tradition as the Georgian pots. If Georgian styling emerged from the possibility offered by handcraft manufacture, the styling of the vacuum cleaner is also a product of the possibilities and limits of mass production: plastic formed by injection moulding. Such methods will tend to result in self-coloured parts that have rounded edges and have gently convex surfaces. If the Georgian craftsman was well versed in the decorative norms of candlesticks, teapots and urns, the modern designer should be equally sensitive to the quality of the surfaces of a product such as the vacuum cleaner. They should also be aware of the way the form and its means of production are related. The vacuum cleaner has been designed in such a way

- to avoid tool-split lines on visible surfaces,
- that the colour breakup corresponds to the parts involved, and
- the part splits correspond to the order in which the device is assembled on the production line.

Thus, the vacuum cleaner's style is as much bound by a form of tradition as the handmade silver Georgian items.

2.5 THE RELEVANCE OF FUNCTIONALIST PRINCIPLES

Functionalism was an attempt to resolve design constraints in such a way as to favour utility over styling. The aim was the search for objectivity, inspired by scientific ideals. However, designers bring their own tastes to the design process and must also show regard to the general professional norms. Because of this, form-giving is not wholly an open question at the start of the design process. It could be said that each designer has a suite of preferred forms waiting to be applied to a given brief. Inasmuch as there are certain formal norms for industrial design, this is the case. These norms, which have some basis in functionalism, have to be brought into alignment with the other factors of quality, cost and, critically, actual functionality. It also matters whether the product looks appealing and also can be distinguished from the alternatives (past and present).

The relevance for form of the debate on functionalism is related to the choices and values of the designer. In the end, you, the designer, have to decide what type and degree of expression is appropriate for the task in hand. Even with clear directions from the customer brief and the information on user preferences, the designer must exercise their own judgement. Often such directions and information are not explicit enough. So, the designer needs a considered rationale concerning what one could call styling.

A short précis of the modernist argument concerning decorative elements is that modernist designers wished to achieve what one could call *objectively* good design. They wished to strip out what they considered unnecessary decorative elements in designed objects. By using all the available information from the science of the day it was thought that the designer could discover the best possible form for an object. In this way personal preferences and traditional decorative motifs could be avoided, leading to the most efficient result: efficient in terms of production, appearance and utility.

There was also a moral argument put forward (in a lecture in 1908, published in 1913) by the Austrian architect Adolf Loos that ornament was a form of crime. As example of what Loos had in mind and wished to avoid is shown in Figure 2.15. Loos contended that decoration was a form of barbarism. Two

2.15
Victorian interior. (Image: Wiki Commons)

citations stand out for our attention: 'The tattooed who are not in prison are latent criminals or degenerate aristocrats. If someone who is tattooed dies at liberty, it means he has died a few years before committing murder' (Loos, 1913). The other quote is: 'No ornament can any longer be made today by anyone who lives on our cultural level. ... Freedom from ornament is a sign of spiritual strength' (Loos, 1913). Loos's contention was that ornament was the design equivalent of tattoos. That is quite some claim given the role of ornament in architecture and design and the prevalent human wish to make things look at least pleasant. While Loos's moral arguments today would not carry much weight, the effect of his arguments, an aversion to ornament (loosely defined), is still with us.

The positive and productive aspects of this anti-ornamental approach are that designs are more suitable for mass production. In the 1900s many of the traditional motifs of household objects were derived from and conceived within a context of handcraft. With the limited and limiting constraints of mass production, complex three-dimensional mouldings and hand-applied colour patterns were not feasible. Mass production favoured fewer parts, simpler processes (stamping, cutting, spray painting). The modernist approach also drew attention to the need for objective validation of design choices, what we consider today to be data such as information on user preferences and ergonomics. Finally, there was a worthwhile wish to make good quality consumer goods more affordable. Up to this point in time, 'designed' goods were relatively expensive and there was a very sharp distinction between what richer people could afford and what most of the rest of the population could pay for.

Jan Michl (1995) summarised these points in an essay called 'Form Follows What?' and provided a critique of the central argument for severe functionalism. His point was that in the end, the designer is *still* the one who decides how an object will appear. Fixed and unavoidable as the laws of nature are, there is still considerable latitude to give a distinctive form to an object. Consider the vast diversity of the forms of cutlery. Even applying all that is known about the series production of inexpensive steel cutlery and taking the functional needs into account, cutlery comes in a huge diversity of equally functional (as in usable, effective) forms. The same is true for clothing. Most trousers do the two jobs of covering and protecting legs entirely adequately. There is no single objectively correct design for trousers. Michl argued that an attempt to pass off a design as the necessary outcome of the laws of physics and engineering requirements is an attempt by the designer to dodge the responsibility for the design outcome. It is also an attempt to claim that a design choice is the sole correct and objective result of a dispassionate assessment of the design problem. In making these claims the designer is seeking to present one design as factually, objectively and absolutely better than other possible designs. In this way the 1920s modernists sought a crushing and unanswerable argument as to why their work was superior to the traditional alternatives. Michl argued that this claim is ultimately unjustified: designers decide whether or not to apply more or less weight to objective factors like efficiency versus the wish for some form of expression or styling. The laws of physics are there but are not the only factors.

To leave as much as possible to engineering and physical laws or to decide to give some form of styling character to an object are ultimately arbitrary choices in the power of a designer. Further, even what appear to be functionalist designs is the result of a great deal of aesthetic refinement. David Pye (1978, p. 11) referred to this as 'useless work'. He did not mean that it is not valuable but that it is not ultimately necessary in the very strict sense. His examples of 'useless work' are the totally flat and clean surfaces of walls or indeed any finishing work at all. By this yardstick much of what concerns industrial designers is 'useless', though this does not mean it is not important or without effect.

While the Bauhaus School of Design set great store on design minimalism and efficiency, even the buildings in which these theories were taught were finished to a high standard, consisting of smooth and neat work organised to a very high degree. This degree of finishing serves no functional purpose (very narrowly conceived), yet it is done anyway. If we look at again at Gerd A. Müller's KM3 Food Processor (1957) blender (Chapter 1, Figure 1.28) we see a design that is incredibly refined and attractive. It is starkly functionalist but also is not the most basic, cheapest device for mixing. So, even functionalist design is larded with 'superfluous' finishing that is as much the result of a designer's taste and customary expectations as the ornate chair in Figure 2.10.

The synthesis of these opposing views is that designers can work according to rules affected by physical limits and customer demands *or* they can design according to their own taste or some balance of the two. They must understand and accept that fact and its consequences when making choices about the form of an object (see Figure 2.16).

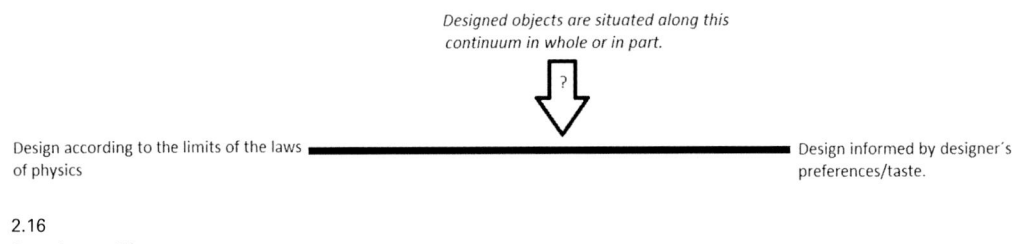

Designed objects are situated along this continuum in whole or in part.

Design according to the limits of the laws of physics

Design informed by designer's preferences/taste.

2.16
Function and form spectrum.

SUMMARY

The aim of this chapter was to provide an overarching set of reasons as to how to select among a variety of geometrically acceptable forms. It is insufficient to be able to generate good quality lines and forms without being able to find the one most suited to its context. Design necessitates compromise, and sometimes this means choosing one form over another for reasons beyond geometrical correctness: cost, quality and functionality. Under quality, the matter of material and production necessities were considered. No surface or line can be selected without due attention to what the object will be made of and how. Finally, the chapter discussed the role of the underlying preference for functionalist solutions and served to remind the designer that a functionalist appearance is also a matter of choice and not only the laws of physics. Designers make choices and need to account for them.

NOTES

1 Successful designs are ones that balance failure the best, as in accommodate the limits of cost and quality in a satisfactory way.
2 Designers Anthony Dunne and Fiona Raby takes a critical approach to design. Their work uses design fiction and speculative design proposals to makes us ask questions of society and design culture. Their work is such that design is in the service of art.
3 Constraints or requirements are the set of factors that are important in a project.
4 Moulded parts in plastic or metal must be able to be removed from the tool they formed in. The draft angle is the degree to which the tool's moulding surfaces are open, usually somewhere between one and seven degrees, depending on the depth of the part.

FURTHER READING

Guidot, R. (2006) *Industrial Design Techniques and Materials*. Paris: Flammarion.
Michl, J. (1995) Form follows what? 1:50 – *Magazine of the Faculty of Architecture & Town Planning* [Technion, Israel Institute of Technology, Haifa], 10, 31–20.
Nygaard-Folkmann, M. (2009) Evaluating aesthetics in design: A phenomenological approach. *Design Issues*, 26(1), 40–53. https://doi.org/10.1162/desi.2010.26.1.40.
Pye, D. (1968) *The Nature and Art of Workmanship*. Cambridge, UK: Cambridge University Press.
Pye, D. (1978) *The Nature and Aesthetics of Design*. London: Herbert Press.
Van Bezooyen, A. (2014) Materials driven design. In Karana, E., Pedgley, O., & Rognoli, V. (Eds.). *Materials Experience: Fundamentals of Materials and Design* (pp. 277–286). Oxford, UK: Butterworth-Heinemann.

3 Lines, surfaces and curvature

3.0 INTRODUCTION

In the following pages we turn our attention to the theory of lines and surfaces. These are to form what notes and melody are to music. And just as musicians can better express their artistic vision by understanding theory of music, designers can do the same by having more insight about their working materials. More understanding leads to more control.

First, I start with asking why curvature quality matters and then move on to how we can discuss curvature in verbal and mathematical terms. We then shall continue by examining the geometry of lines and follow by considering curvature in mathematical terms. Curvature is dealt with under the heading of line character (using terms designers will use in relation to CAD programs). That leads to how to think about blends from one main surface to another, or what is called curvature continuity: positional, tangent and curvature continuity. Good curvature continuity supports forms whose surfaces will have harmonious reflections. Reflections are what we see when light bounces off an object. We also need to consider shadow, darkness and contrast. Positive and negative space is discussed in terms of the interaction of light and shade. The chapter concludes with a short discussion of the industrial design corner and edge, the lines between different parts or panels and, finally, applied graphics.

This chapter presents:

- why curvature quality matter blends from one main surface to another (transitional surfaces)
- positive and negative space in relation to interaction of light and shade
- the industrial design corner and edge
- the lines between different parts or panels
- applied graphics.

Surfaces are considered in greater detail in this chapter. Rather than being a field of unlimited variables, surfaces exist in a limited continuum from flat to curved. In other words, the potential range of curvature is presented as a

DOI: 10.4324/9781003183303-3

variable the designer will actively control and not some fluffy, vague mystery. This is with respect to the meaning of the curvature quality (just as designers seek to control functionality, colour or material). The way surfaces of different curvature character can be combined is where there is considerable latitude for variation and invention. Understanding curvature is essential to being able to construct good quality forms, and this section will go into practical detail on what curvature quality means in technical terms.

Being able to judge the quality of the three-dimensional object means assessing whether the visible lines (edges, joints, highlights and shadow boundaries) are good in themselves but also whether they are sufficiently close to the design intent. This chapter provides a means to do this by explaining the underlying mathematical quality of lines (and, by extension, surfaces) and so shows how to get a more objective view of the design's character.

Understanding the character of a line involves being able to read its curvature and to extract the signal from the noise. The noise might be surface irregularities, unwanted shadow or optical effects.

One way the following sections can be related to each other is shown in Figure 3.1. Follow the arrows to make sense of it.

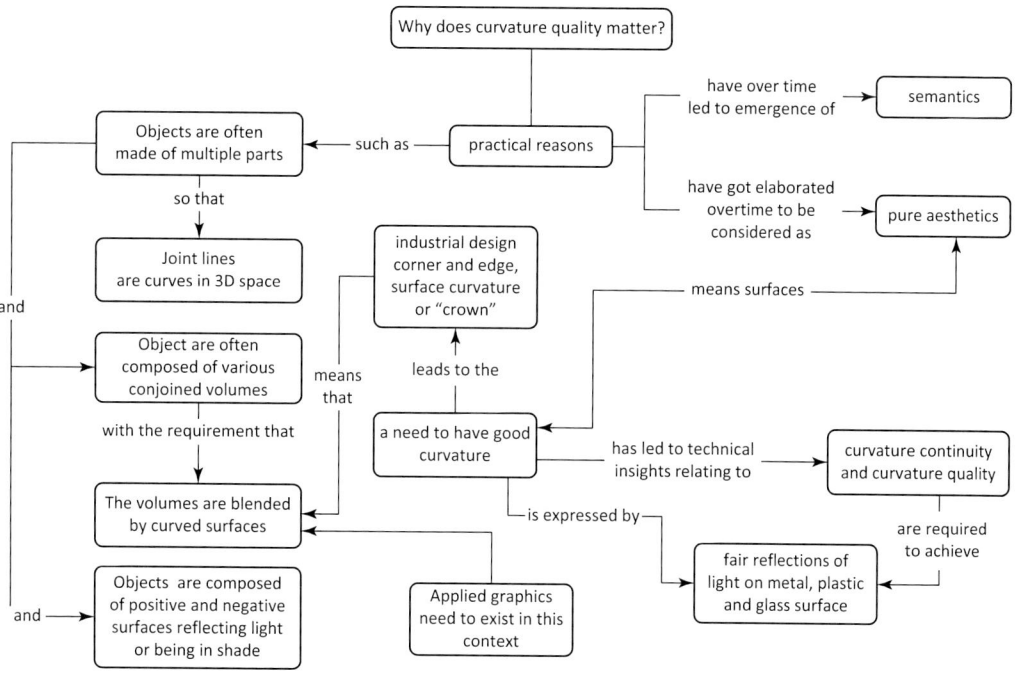

3.1
Mind map of the contents of this chapter.

3.1 WHY CURVATURE QUALITY MATTERS

Knowing why line quality and surface quality matter (which is what I argue here) may make the more abstract sections which follow a bit more enticing. So before addressing (lightly) the geometry of curvature, some reasons are needed to show something of why we might call one set of lines pleasant and another less so.

There are two arguments for why aesthetics of curvature matter; the parable later in this section is an attempt to first identify these reasons in a context other than that of three-dimensional form and design.

Argument one is that people generally prefer smooth transitions to abrupt ones. Argument two requires something of a metaphor based on the simple notion that people mostly prefer to have their wishes met and among those wishes is to be treated well.

Addressing argument one: industrial designed objects are typically closed or semi-closed compound surfaces, all connected together. Take a look at the casing of a lawn mower, a car body, the seat and back of a plastic office chair or the sculpting on a shampoo bottle. Many curves and surfaces of these things are transitional in nature (see Figure 3.2), forming a blend from one area to another. Generally, smooth changes from one section to the next are more pleasing for the same reasons we like smoothness elsewhere in life. Note that I am not saying that all forms must be smooth. The point is that where smooth transitions are required we want them to have controlled changes of curvature which make for good reflections.

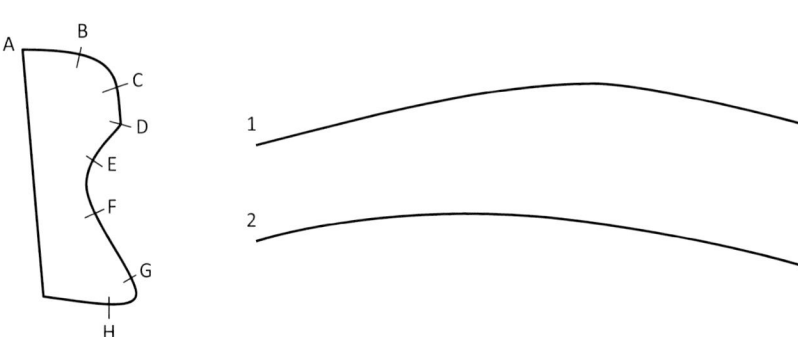

3.2
There are mostly smooth transitions between sections of the curve A to H. But at D there is an abrupt change from one curve section to the next. Curve 1 shows a marked bump. Curve 2 shows a flat area in the middle. It looks ambiguous.

This point about curvature was made by Gomez-Puerto et al. (2016), who wrote: 'That people find curved contours and lines more pleasurable than straight ones is a recurrent observation on aesthetic literature'. Curved lines 'have often been regarded as more harmonious, relaxing or pleasant – and more in consonance with nature - than straight or broken lines' (Gomez-Puerto

3.3
Snake by Phil Price, Aarhus, Denmark. Note the smooth transitions between sections of the work. The aesthetic effect of the structure depends almost very much on the gentle curves being smoothly blended.

et al., 2016). Bertamini et al. (2015) cited the painter Hogarth, who talked of curved lines as expressions of grace and beauty, the so-called line of beauty. Figure 3.3 shows a building whose architectural merit depends on four curves being suitably managed.

Now it's time for argument two, which relates to wishing to be cared for and treated well. For this we need a parable or metaphor.

Consider a waiter who carefully sets down a chair for a guest and then quietly places cutlery on a table and neatly pours some wine. The aim of the waiter is to do the work unobtrusively. The waiter also wishes to acknowledge the innate worth of the customer as a person: basic good manners mean things are done with care. As a counterexample, a waiter who lets the chair clatter heavily on the floor, noisily puts the cutlery down and spills the wine would certainly disturb the customer at hand and anyone else nearby. The signal sent out by these sorts of actions would be that the waiter lacked competence or, more likely, simply didn't care much about the customer or providing good service. The meaning of the smooth service of the waiter derives from understanding the care and effort demanded by this mode of behaviour. Slowly lowering a chair takes effort, time and control. To carefully place a spoon or fill a glass requires judgement. In these examples the transitions are spatial: a chair is moved, a spoon is placed, a glass is filled slowly. By analogy, well-judged forms (smooth transitions, good surfaces) are the design equivalent to showing the care and courtesy that the considerate waiter demonstrated.[1]

Moving on to other types of transitions, again there is a general tendency for people to prefer gradual changes in space and in time. In visual-spatial terms, one of the appealing aspects of a classical town centre is the way in which the elements relate to one another, chiefly through the use of a limited set of materials and building types (Weber, 1995). The spatial changes of a classical city usually have gentle transitions and are thus easy on the eye. When something breaks the expected arrangement of masses such as a sudden vista on a square or the dramatic spectacle of a church, it is usually a pleasant effect.

In terms of change over time, it is considered more pleasant to gradually change light levels from dark to light or vice versa. It helps the eyes adjust and is less confusing. Musical compositions also change over time and feature the careful use of small steps from quiet to loud or from fast to slow or perhaps the use of expected versus unexpected sequences of notes. Sudden changes can be used to good effect, though. When a composer wishes to make an impact, they often choose to do so by the use of sudden changes of rhythm key or volume, sharp contrasts and the use of surprise. However, even if not all sudden changes in state are unpleasant, often unpleasant things are abrupt either spatially or temporally.

Shock is sudden. Surprise and shock are closely related.

EXERCISE 3.1

The aim of this exercise is to make you aware of the way shapes are or are not smoothly blended.

Considering transitions and changes, create a list of changes and class them according to the degree of pleasantness of unpleasantness; for example, being splashed by cold water, the stage curtains opening at a theatre, the start of a piece of recorded music, the sensation of eating a spicy food item, the way buildings join each other on a street.

Rank the speed of the change (relative terms will do). Try to look at the relationship of the transition to its context. Consider the meaning of the transition as in what you think the artist/designer had in mind, where relevant. In the case of music, how do the parts of the composition blend – sudden or smooth? How does a building fit into its context? Are the parts of a consumer product blended or merely abutting?

The other angle to this preference for smooth and controlled transitions is to do with effort and the signal it sends to other people. As mentioned in the parable, the careless waiter's behaviour signalled a lack of respect. It wasn't smooth or controlled. Little effort was expended to show that the customer mattered. By analogy, an object with carefully managed forms is an object that signals respect: someone cared to make an effort on behalf of the user. A case could be made that the joyous aspect of good civic architecture is that great effort has been made on behalf of the citizens who will use it. A great product is one where the user recognizes that much energy and thought has been given to the user's experience of it. Many users can intuit the higher value of well-modelled objects (Ming et al., 2001; Westermann et al., 2012).

Figure 3.4 shows a very blunt demonstration of the impact of curvature quality. On the left is a mug which was bought new for £1.70 (adjusted for inflation)

3.4
Staffordshire Tableware mug (left); Dibbern mug (batch 6205) (right). (Author's collection)

and on the right is a mug bought new for £19 (adjusted for inflation). Both mugs are functionally identical, but one, through the careful elaboration of its form, can command ten times the price of the other. More significant than the price of the object is the way the Schoenwalder (right) seems like a mug for life, an object that can become part of the fabric of daily routine, adding grace to ordinary moments.

It may seem like a long way from the grace embodied by a ceramic mug to considerations of geometry, but much of the Dibbern mug's value and appeal lies in the correct and sufficient use of curvature to convey value. The mug has a surplus of meaning, to use Nygaard Folkmann's (2013) concept, whereas the Staffordshire mug carries almost none; it is just a simple mug. Any significance it has is negative: here is a mug made with no further intentions; its statement is to make no emphatic statement. It means either one doesn't care about the object or that one wants to show one does not care about such things. There is a place for such functional design, but it must be used consciously. And when one wants to rise above plain simplicity, which becomes monotonous, one must be aware of the factors involved in dealing with expressive curvature.

EXERCISE 3.2

Try to elaborate on the underlying reasons why the Staffordshire mug could be considered less appealing than the Dibbern. Where in the mugs' characters is the nature of the difference? You might want to look into some of techniques of ceramic manufacture to work out the answer.

Next, find two objects of the same type and examine how the different solutions for the form-giving were approached. This means a detailed inspection of the shapes and preferably objects with a decent amount of curvature. Kettles are a good example.

Ideally, if you are working in a class, work in teams of two. Person A finds a pair of objects (e.g., two stackable chairs) and Person B finds a pair (e.g., two table lamps). A and B swap objects and try to assess which one was the better and the worse of the two. Then explain your reasoning. Did A agree with B and did B agree with A about the assessment? You'll need to take a very critical approach to these objects since the differences may be small. Here we are interested in surfaces, curvature, smoothness and refinement.

3.2 HOW TO DISCUSS CURVATURE, FROM VERBAL TO MATHEMATICAL DESCRIPTION

To describe form we can use language to some extent, deploying figurative and metaphorical terms. We can understand and express qualitative attributes in this way. However, language is limited. There are only a few broad terms in English that deal with curvature: straight, relatively straight, slightly curved, uneven, accelerating, asymmetrical or snakelike, weak, muscular and so on. You may find you wanted to either draw the concept you are trying to explain or wish to gesticulate; that is, to use hand motions to express the idea. So, this vocabulary is not sufficient to discuss the curve (or indeed surface) in more than a general way. We wish to go further, to start to express the curve quality precisely. With a lack of expression there is a lack of control. Thus, we need some help from science, in particular, geometry.

EXERCISE 3.3

This exercise will take a while, but the result is that you can see how little words can convey or how long it takes to get an idea across with the few terms we have.

Find an object with some interesting shapes and either bring it to class (hidden!) or take some photos. Working in pairs, student A asks a fellow student B to draw an object selected by student A; for example, student B draws the alarm clock that student A has brought along. The instructions are only verbal. The tasks stops when time runs out or student A thinks B's drawing is close enough to the original.

The next step is for student A to show the object to student B and to compare the drawing and to consider especially the surface treatments. Did student B's drawing get close to capturing the shape of the object student A had in mind?

With this in mind, we see that we need a way to think about the bridge from vague verbal concept to definitive geometry. We start from an emotional, verbal concept and produce a drawing; the drawing yields a model; the model needs to be controlled so that the curves and proportions are in accord with the theme sketch and the curve quality is good enough and true to the initial emotional, verbal notion.

We can see that the curvature quality of a line has a mathematical basis in that it can be defined by an equation. Whether the line is appealing or not depends on common human preferences and values. That is, the quantitative aspect of the line has a qualitative effect just as 5 mg of salt (an objective fact) has a subjective saltiness to it. Most people will agree that 5 mg of salt will be sensed as salty. Numbers can be put on curves, too, which can help in controlling them (making them how we most want them to be). The point here for the design practitioner is that if they are to have any hope of managing the subjective aspects of form, it's good to have a basic grasp of the quantitative

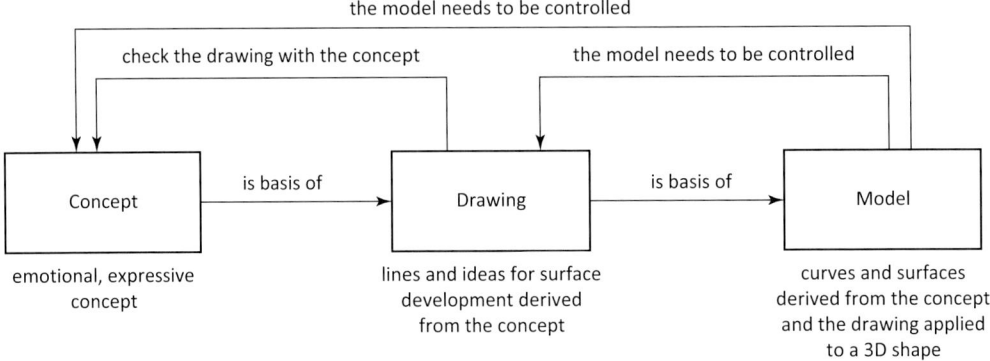

3.5
The feedback from the concept to the drawing to the model. Puzzles at the drawing stage can be related back to the concept, which can be revised, and puzzles in the model stage can be worked over using new drawings and revising the concept until all three are in equilibrium.

or measurable aspects influencing that form. If so, the designer can specify a curve with a specific mathematical description and must also have a view on how it is perceived subjectively.

Table 3.1 shows how the qualitative and quantitative are related.

Table 3.1 The quantitative part is basic but might only be determined later, when the saltiness, volume or curvature come up for discussion. The intersubjective part is what most people would agree on; the subjective part is where judgement and taste come in.

Parameter	Quantitative	Intersubjective	Subjective
Salt content	5 mg/1000 g	"It's salty"	"Not salty enough"
Volume	92 dB	"I hear a sound"	"Too loud"
Curvature of a line	10 mm	"There's a radius"	"Too rounded"

The implications of this are that there is a number attached to most qualities. A certain level of agreement can be reached about these, and this is called intersubjective agreement. In the examples in Table 3.1 the category 'intersubjective' covers things most people would agree was the case; for example, this has salt in it, there was a sound, that thing is curved. The subjective part is where judgement comes in and the designer has only experience and taste to guide them. If the person who judges the curvature to be excessive knows it has a radius of 10 mm, they can try reducing it to 8 mm to see if it looks better. If they don't have a handle on that number they may end up making random trials or not even know where precisely the problem may lie.

The aim of this section was to show that curvature is a measurable quality like other more familiar ones. It is related ultimately to subjective perceptions. To influence those subjective perceptions, we need to dive further into curves and geometry.

EXERCISE 3.4

The aim of this exercise is to show the looseness of the relationship between words and images, especially curvature.

What to do: 1) write a list of verbal expressions describing various degrees curvature. Then 2) generate a set of visual images that exemplify these terms. You could draw them or hunt for existing images. 3) Match the images to the word by putting a sticky note on the drawing. More than one word-to-image match is possible. You may consider the whole object or part of it.

Notice that the objects may fall into one or more word categories and that an object as a whole may not have the same curvature character as smaller parts of it. Why might this be the case?

You could work as a class. Put unlabelled drawings on the wall and, as a class, place sticky notes on the images. If you have the word 'wobbly' on a note, attach it to a drawing you think has a wobbly element, for example. Are there labels everyone agrees on? Are there ones that generate disagreement?

Here are some suggestions for words related to aspects of curvature: serpentine, arciform, angular, flat, exciting, calm, elliptical, crooked, wobbly, arched, dynamic, flowing, truncated, unsettling, swirly, snaky, circular, smooth.

3.3 THE HIERARCHY OF LINES, CURVES AND SURFACES

At a conceptual level, designed objects can be understood as being composed of an infinite number of lines which seem to change form in three-dimensional space. Designers use drawing as a tool to economically show the lines that define the essential character of the object. The drawings thus capture the primary lines which describe the object, or the design intent. Shadows and reflections can be thought of as linear features because shadows have edges, too.

Lines form the basis of how form is understood and represented. They also show how forms are constructed in three-dimensional hard models and virtual, CAD models. In essence, lines are the shorthand of designers but are also real entities in the world. Unlike language, the relation of the notation (a line on a page) to the thing it denotes (the object) is direct. Human language is, in contrast, a convention of matching arbitrary sounds to objects and ideas. The 'houseness' of a house is not contained in the word or utterance of 'house'. In drawing, the character of an object is manifested in the lines representing it as well as the thing itself.

Rather than being a field of unlimited variables, lines exist in a limited continuum from straight to very curved (spiral). By analogy, surfaces vary from flat to spherical. Manipulating form involves being able to conceive of, critique and control the variations in curvature of simple and compound geometries. It also requires sensitivity to the relation between the curves and surfaces of which an object is composed. That makes it relational. There are the relations of the curves *within* the object and the object as it relates to *others* of if kind (see Figure 3.6 for the levels of analysis).

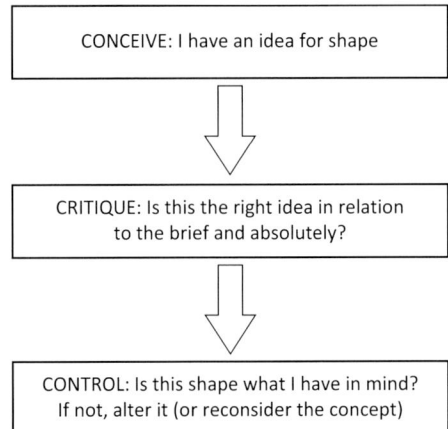

3.6
Conceive, critique and control.

The hierarchy of lines and surfaces can be arranged in two almost entirely conflicting ways. Typically, the order in which a product is drawn has nothing to do with the way the geometry would be built up when modelling it. A product can be defined in almost any order, depending on what the designer is interested in: outline, colour break-up, proportions, joint lines or even shading or highlights. Assuming there are some detailed drawings of the object to start with, the sequence of construction of a model (CAD or hard model) is usually much less variable and probably starts with a definition of the outline and major features using intersecting planes and surfaces as construction tools. What this means for us is that there are two ways to create a hierarchy of lines and surfaces: as they are drawn or as they would be built up in 3D.

The following sections are sequenced according to the geometrical hierarchy of points, curves and surfaces. Joint lines and applied graphics are discussed in Sections 3.10 and 3.11, respectively.

EXERCISE 3.5

The aim of this exercise is to show that you can begin drawing a shape using any element as a starting point and that the results will differ depending on which element is chosen as first one. You can also use this method to break your drawing habits. If you always start with an outline, then try choosing a different element instead. You may be surprised as to what the result looks like.

Choose an object category such as domestic appliance, motor car, shoe or radio, for example, or even a building and begin to draw a new shape based on *one* of these elements first:

*   Overall outline
*   Proportions
*   Colour break-up
*   Joint lines
*   Shading and highlights
*   CAD construction principles[2] (if you are familiar with them).

Having established the first element –for example, shading and highlights – now add the rest. If you are working as a class, assign one category to each student. Be sure to retain a copy of the first element to compare to the finalised sketch. This allows cross-comparison between sketches by different groups. See what differences there might be in the resultant drawings. You can use this method to build a drawing from different starting point. If I draw graphics first, the profile is dependent on them; or if I draw proportions first, the other features will be controlled. A highlights-first drawing usually results in a very organic, smooth shape.

3.4 LINES IN THREE-DIMENSIONAL SPACE

It is time to move on to some geometry to before delving into to surfaces and surface quality. It is true that designers seldom intuitively think about lines from the point of view of geometry. However, eventually the geometry of the line will need to be dealt with in order to better manage the surface quality and the resultant reflections and shadows. Simply put, if you understand some basic geometry, you will be able to deal with what's working and what looks goofy or strange.

It is a curious aspect of design that the fairly simple geometrical principles related to understanding form are overlooked or even resented. The most basic element of design is a line which is a geometrical entity. In music it is the note. Unlike musicians, who must understand their own class of technical issues like pitch, timbre and duration, designers seem not to be interested in the mechanism of design's essential matter. This section aims to show that to be fully in control of the aesthetic perception of a form, a designer must understand the geometry that is the first cause of the aesthetic effect.

Starting from basic geometry, points exist in space and have a location in the X, Y and Z dimensions.

Conventionally, the dimension X represents length, Y represents depth (or width) and Z represents the vertical height of the object when it's being modelled (see Figure 3.7).

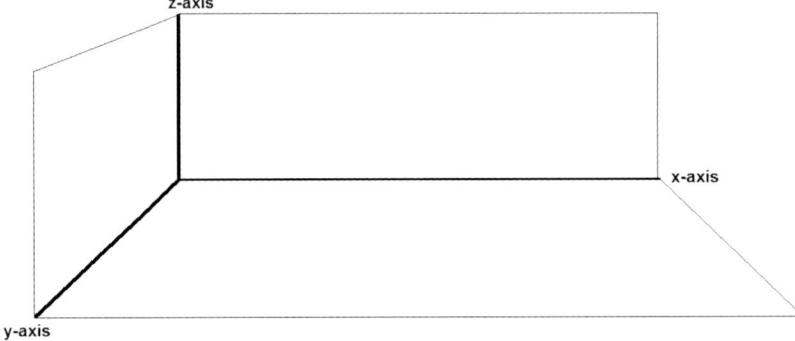

3.7
Axial space. Z is up, X is length and Y is width.

Two points, A and B, separated in space, can be connected to form a straight line (Figure 3.8).

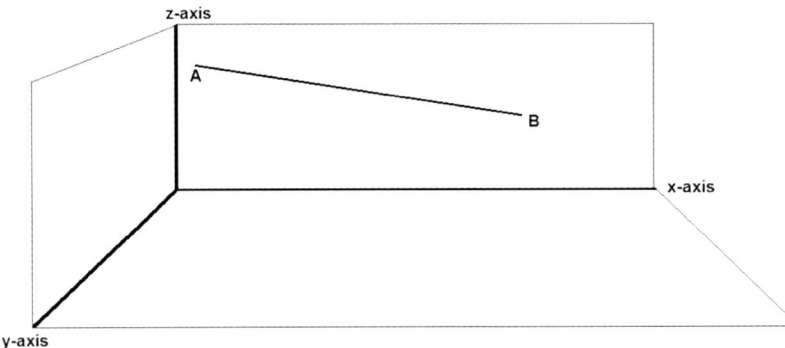

3.8
The straight line in space. It is situated on the plane of Y where Y equals zero and has a height difference on the Z axis (A is higher than B). B is further away from the origin 0,0,0 where X and Y and Z are all 0.

88 □

By adding a third point, C, and joining them up, one of many potential curves can be formed. Figure 3.9 shows one possible way to join A, B and C.

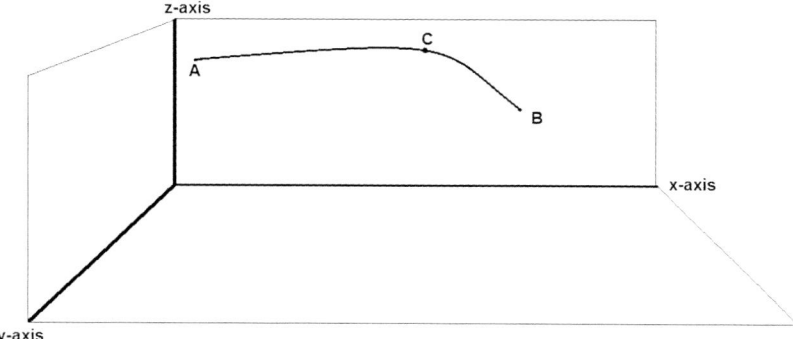

3.9
A curved line con-
necting points A, C
and B.

If the middle point is the same distance from both ends (equidistant), the curve could consist of an arc which is a section of a circle (Figure 3.10).

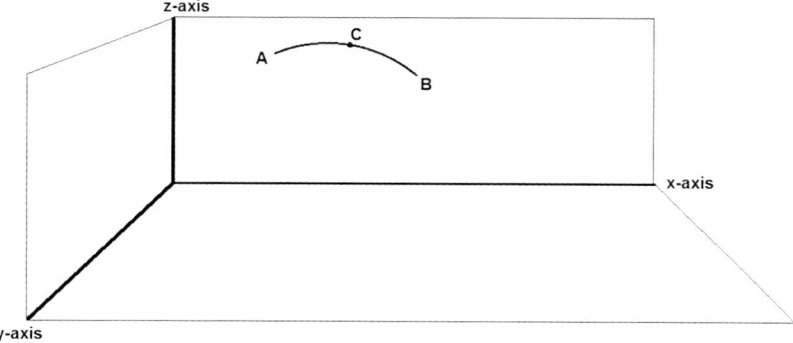

3.10
The arc in space. It
is still two-dimen-
sional. The 'snake' in
Figure 3.3 is strongly
three-dimensional.

Other solutions exist with three points, but the arc is the simplest involving one line which is a curve.

If the third point is not equidistant, the most likely solution will not form an arc but a curve. It could also have been an arc, because the three points lie quite close together on the Z axis (Figure 3.11).

The character of these curves can be described in the terms discussed in Chapter 1, Section 1.6; that is, in terms of forces acting on the curve to deform it from the straightest path. We can also say the curvature is greatest using the co-ordinate system to numerically define the precise point.

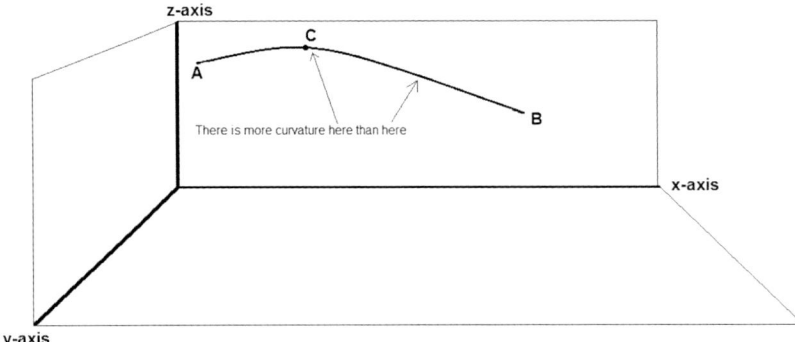

3.11
The arc in space. It is still two-dimensional. The curve shows more activity near the 'C' than closer to the 'B', where it is flatter or less active.

Curves are used to define surfaces. Surfaces, being three-dimensional, are defined by three, usually four, lines making up the edges (see Figure 3.12). What is true of curves applies to surfaces. The language of seeing 'as if' can be supplemented with precise co-ordinate descriptions of where the surface is in space.

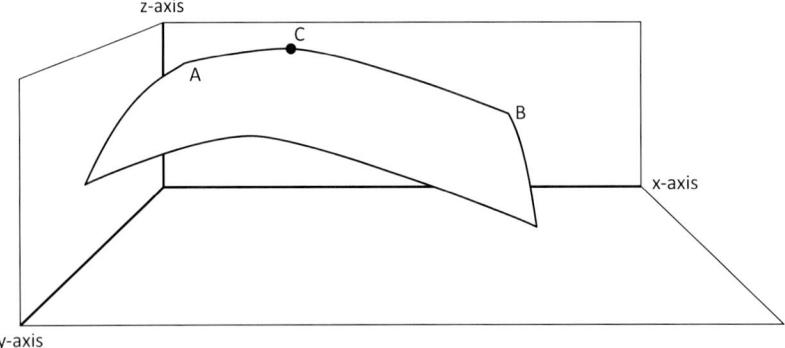

3.12
The curve has now been extended into three-dimensional space. It is now a surface with four edges. There is a gestalt principle operating in this image now. What is it? Continuity. The Z axis and X axis lines are interrupted by the surface, but one assumes they are still there, hidden.

Since designers don't usually think in terms of points in space, this might seem abstract. When it comes time to create models in CAD software, the underlying geometry of curves becomes very relevant. The task at hand is to build up three-dimensional forms from basic curves consisting of, usually, three or more points. What we have in Figure 3.12 is a surface made of four curves defining the edges and influencing the shape of the surface within them. Designed objects are composed of these units, situated in X, Y and Z space. In the next section we will attempt to understand the nature of the curvature of those curves and the resultant surface.

EXERCISE 3.6

The point of this task is to focus your attention on shaping a harmonious line that must pass through several fixed points in an economical way. It is a bit like the job of designing a casing around the mechanical parts inside a product. The different lumps need to be enclosed in a fair-looking shape that is made up of different surfaces.

On a large piece of paper, draw four dots, or more if you wish. Place a piece of tracing paper over the dots and use either careful hand-drawing or flexible tape[3] to connect the points with the smoothest lines possible. See how many plausible ways there are to connect the dots without adding extra 'information'. Extra information would be something like convoluted loops used to join up dots in a roundabout way.

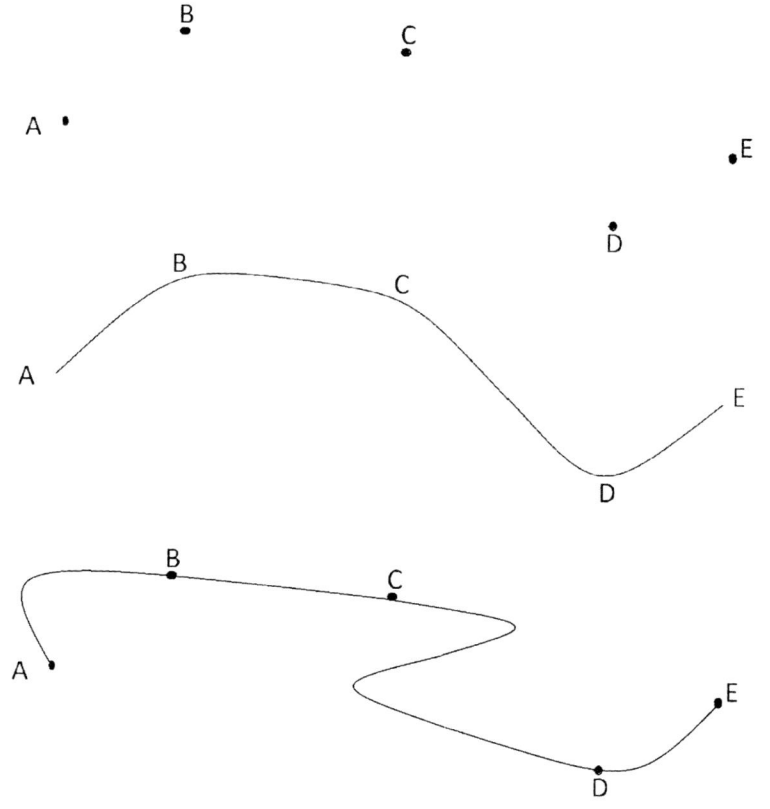

The example shows five dots, A, B, C, D and E, with two solutions. The upper solution joins the dots with less 'information' than the lower solution. In the lower solution the curve seems to head off to a fifth point between A and B; it also has an extra zigzag between C and D.

3.5 CURVATURE

Regardless of whether you will use a CAD programme or not, it will help to have some conception of the concepts related to curvature. This understanding will allow you to describe more precisely the underlying reason for a given aesthetic effect much as musicians have a technical vocabulary for discussing the cause of a given aesthetic effect in a composition or song.

The thing we want to get out of this is what a fair curve and fair surface look like. Consider the drawing in Figure 3.13.

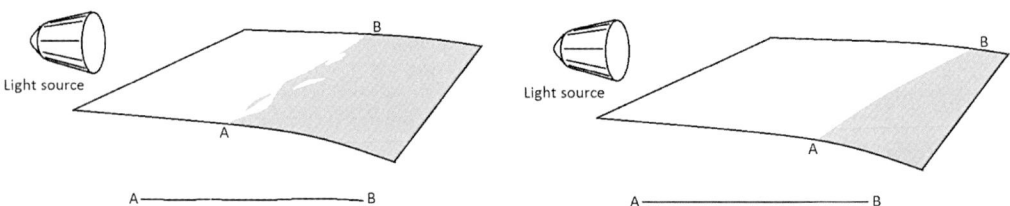

3.13
An uneven surface and a section, A–B, showing the wobbles in the surface (left). An even surface with smooth curvature (right). A section A–B showing a clean, smooth line (bottom).

Since industrial design deals so much with complex shapes that are unified into flowing forms, we are very interested in making them look as refined as possible. So we need some basic concepts of curvature.

First, we need to take a simple case of a curved line.

3.14
A simple arc.

Figure 3.14 shows a simple arc. That means it is has constant curvature. If the ends of the curve are continued (extrapolated) at the same rate of curvature, they will meet and form a circle (Figure 3.15).

The drawing in Figure 3.16 shows a compressed circle.

You will notice that that curvature gradually increases from A to B. It is flatter on the top (and bottom) and most curved at the sides. If this flattened circle was made into a 3D shape, like a squashed ball, it would have evenly changing curvature from the top to the side. Reflections might be distorted but in a gradual and regular way. That would be an example of fair curvature, dynamic but smoothly changing. The word for this change in curvature is 'acceleration'.

From A to B the curvature accelerates, meaning it gets more and more curvy. From B down to the bottom the curvature reduces again. The point of maximum curvature is at B (and the same on the right side of the shape). This then is a simple example of what would be in 3D a curved shape with good

3.15
A circle is a closed
line with constant
curvature.

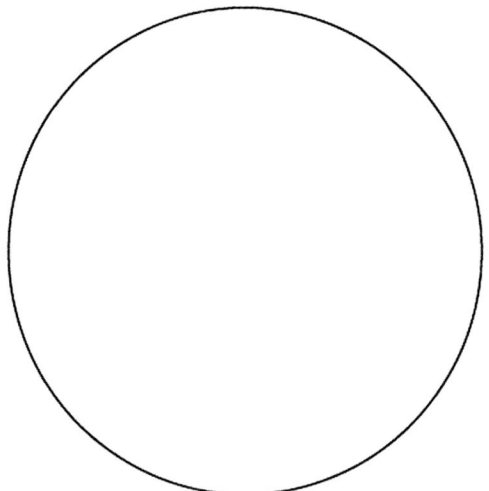

3.16
This looks like a
flattened circle. It
has more curvature
at the sides than at
the top. The rate of
curvature increases
or accelerates from
the top part to the
sides. From the sides
down, the curvature
decreases again.

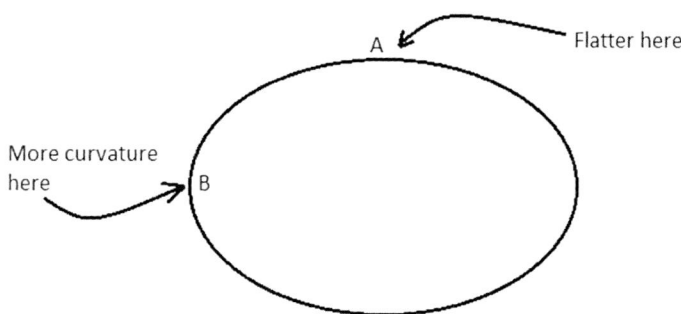

quality curvature. If you look at the work you did in Exercise 3.6, you will have
been trying to make lines with smooth changes in curvature; that is, where the
acceleration from less to more curvature is smoothly increasing or decreasing.

If we look again at the curved surface in Figure 3.12 (redrawn with mark-
up as Figure 3.17), we see that the maximum curvature in the middle of the
surface. The curvature accelerates in the direction of the arrows.

The image in Figure 3.18 shows a three-dimensional curve on a blender.

Notice how the outline of the panel changes in a controlled and consistent way from the centre to the corner marked A–B. It is hard to discern where
precisely the point is where the less curved section becomes the more curved
corner. Although the line shown in Figure 3.16 is three-dimensional, it is still
a line (the edge of the light grey panel). It is not a surface. The condition of
smooth changes in curvature looks like the image in Figure 3.19, which shows
strip lighting reflecting on the body of a car.

What we notice are the smooth lines of the reflections and not the surface
itself. The small irregularities in the edge of the white stripes are due to the
very small ripples in the paintwork.

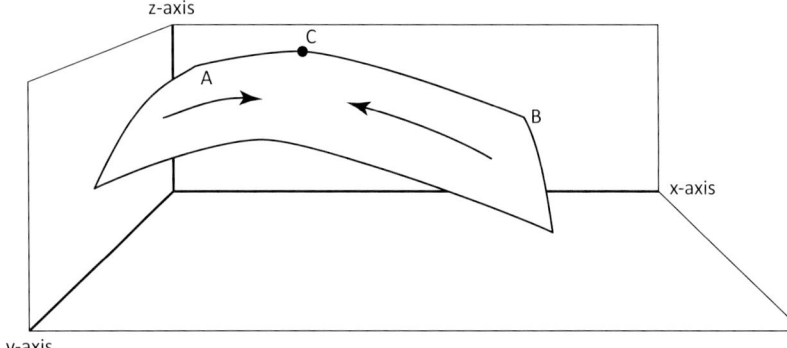

3.17
The arrows show the direction of accelerating curvature towards the middle of the surface. The surface has an axis of curvature parallel to the Y axis.

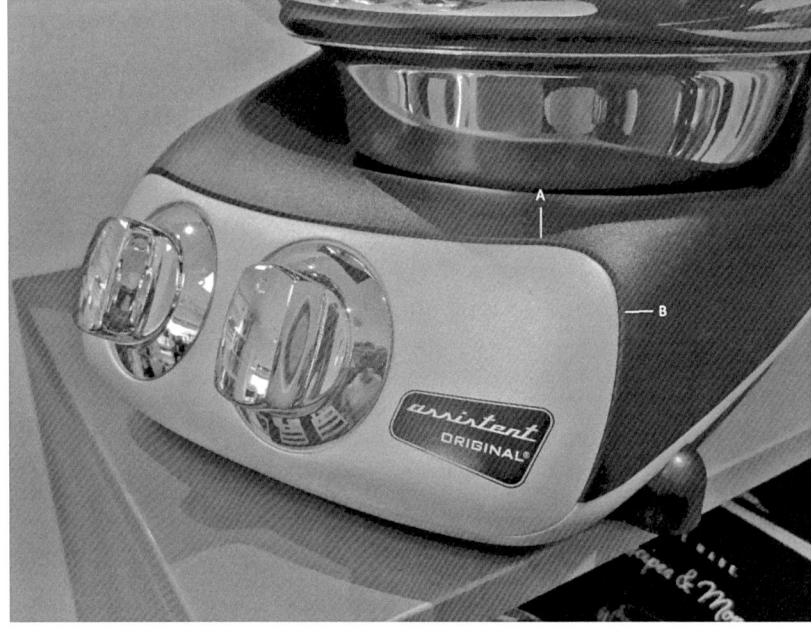

3.18
Gradual acceleration in curvature from the centre to the corner marked A–B. The curvature is most pronounced between A and B and then diminishes as the line goes downwards again.

In the case of this artwork *Waiting Swallow* (Figure 3.20) by Irish sculptor Ian Pollock, curvature continuity is the dominant aesthetic character and lends to the work its characteristic appeal.

Even small changes to the smoothness of *Waiting Swallow* would markedly affect the way it was perceived. Recalling the metaphor of smoothness and palpable care, this artwork's aesthetic merit in large part rests on the congruence between the manifestly evident toil required to make it and simplicity of the bird-like form it represents. The form is an abstraction of a bird and the aesthetic interest then is drawn towards the essence of the form, its continuously flowing surface development. The work's power resides in the subordination of the content to the refinement of the form. The best product design achieves this with the added complexity of multiple parts.

3.19
Highlights from overhead strip lighting on a car bonnet.

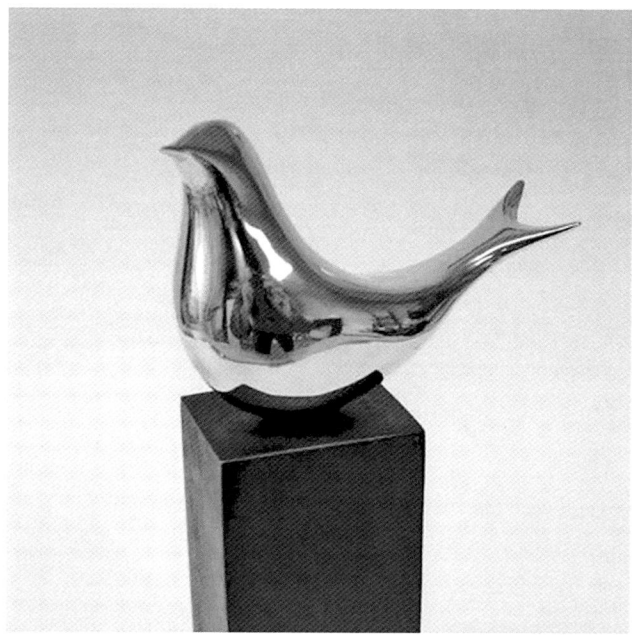

3.20
Waiting Swallow
(2017), Ian Pollock.

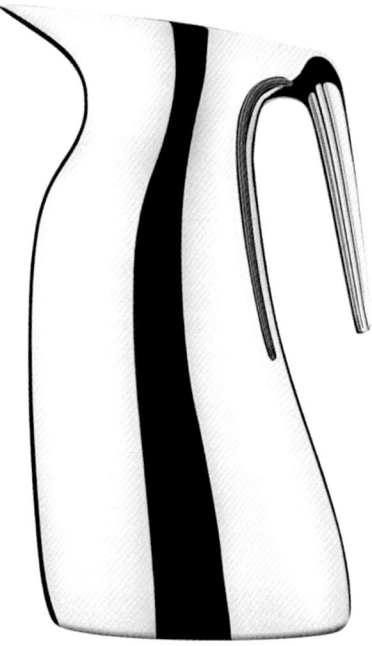

3.21
Beak jug by Georg Jensen, Denmark.

This illustration of the Georg Jensen jug (Figure 3.21) shows how the light flows in a smooth and even way across the surface. The quality of the design and its luxurious character is expressed by the interplay of the highly reflective surface and the environment in which it is placed. This particular detail (Figure 3.22), where the body joins the handle, shows very fine control of the highlights.

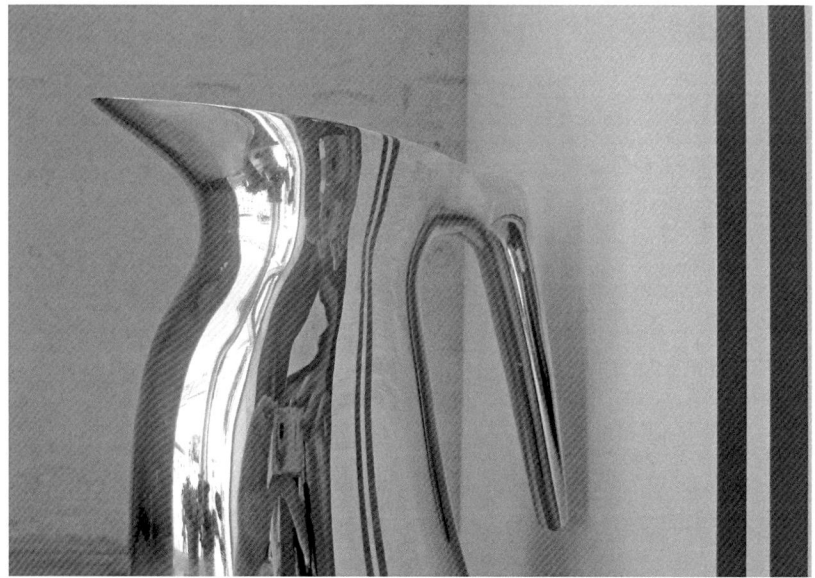

3.22
Beak jug by Georg Jensen, detail of handle and body junction.

If one of the points of fine art is to direct our attention towards pure sensation, these three examples (the car bonnet, the sculpture and the jug) demonstrate this clearly. In each case the expert manipulation of surface curvature and material quality makes us look at the interplay of light and form for its own sake. In the case of *Waiting Swallow* by Ian Pollock and *Beak* by Georg Jensen, all distractions have been removed to leave one central idea in the mind of the viewer. This leads to a phenomenon Immanuel Kant (1928) called the aesthetic moment in which all concerns of utility and meaning are left aside as one is immersed in pure experience.

All of this apparently technical detail is what is behind the pleasing or even very pleasing surfaces that are the hallmark of carefully finished industrial design. Just as a musician is concerned with notes, their volume and speed of delivery, a designer needs to be alert to the character of three-dimensionally curved surfaces. If handled correctly, the effect is a visual symphony of forms that are related harmoniously together and look fair from all angles. This demonstrates, as with the parable of the diligent waiter, that the designer is interested in the user to the highest degree. Of course, fine surfaces can't substitute for adequate functionality. But nor is adequate functionality quite enough. The two go together and are of equal value. This section also shows the bridge from objective geometry through to the highest level of aesthetic experience, the 'wow' effect, or, in Kant's terms, the aesthetic moment. If the design works aesthetically, then for an unspecified time the viewer is unable to think of anything else other than that which confronts them.

EXERCISE 3.7

For this exercise you should be fine-tuning your sight and touch to the fine-scaled nature of curvature and sculpting. Try to see what is there and not the general impression of what ought to be there.

Find a moulded plastic object or something composed of smooth compound forms. Take a photo and, on the photo, mark out the main surfaces and transitions. Annotate the areas of greater and lesser curvature change. What is the quality of the surface like? Are the transitions smooth? Is the object homogenous in that that the treatment of primary and secondary surfaces is the same? Is anything interfering with the perception of the object as whole; for example, graphics, colour break-up or gaps between the parts? Where would the object lie on the scale developed in the first exercise in this chapter? Is it about smoothness or angularity or some combination?

3.6 CURVATURE AND COMPOUND FORMS

The reason the matter of changes of curvature is so important is to do with the fact that the designer has to work with complex shapes like the Georg Jensen jug. Many products of the industrial design process are composed of either refined monovolumes (e.g., some form of rectangular shape) or compound forms where one can discern one or more main forms and subsidiary details such as feature lines or indentations. The car in Figure 3.19 is an obvious example, but I will consider simpler forms to begin with. Figure 3.23 shows a simple fillet where one surface blends into another. The car body shows main surfaces and much smaller ones: the fillets which are transitional surfaces making up what are seen as the edges of the object. Figure 3.24 shows two surfaces bridged by a transitional surface. It's a relative term. If you zoom in close, the small fillets are themselves transitional surfaces.

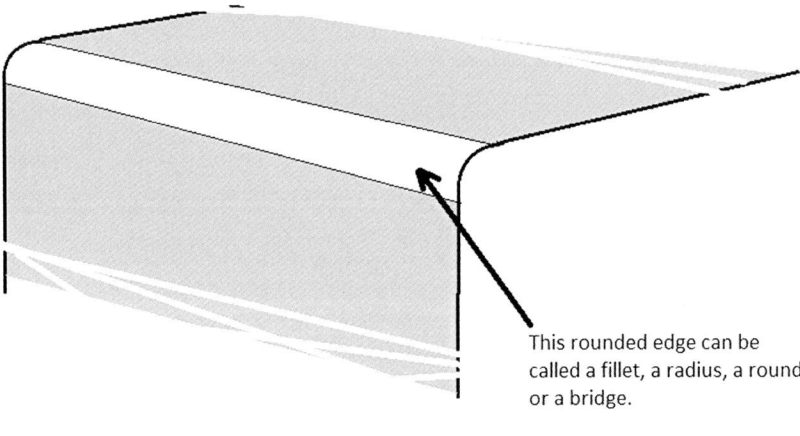

This rounded edge can be called a fillet, a radius, a round or a bridge.

3.23
A rounded edge. Probably a third of the time spent modelling a complex form might be taken up doing rounded edges. It can get tricky where the fillets meet up.

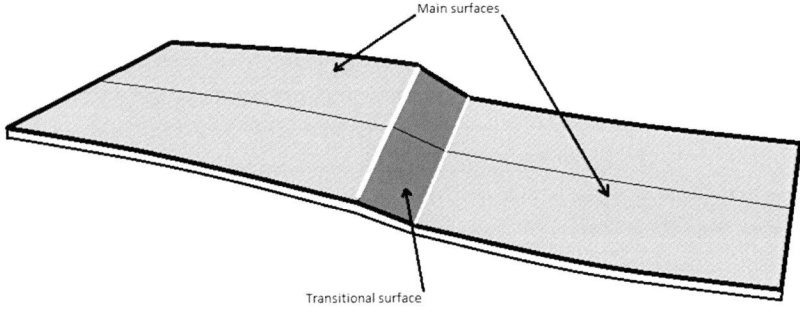

Main surfaces

Transitional surface

3.24
A generalised diagram of two main surfaces and a transitional one.

A basic shape such as the simple rectangular form (Figure 3.25, top left) will usually conform to the industrial design norm of having rounded corners and gently curved main surfaces. The image shows a basic shape with no rounding, for clarity. When finished, the curvature of the main surfaces and of the radiused edges must be satisfactory in themselves.

3.25
A simple rectangular monovolume (top left). A compound form with a primary and secondary volume (bottom left). On the right side are versions with rounded edges added.

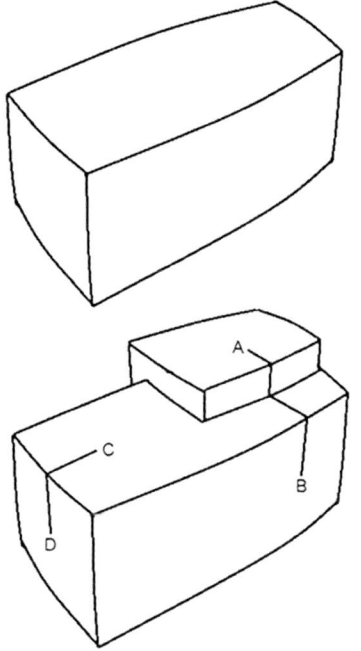

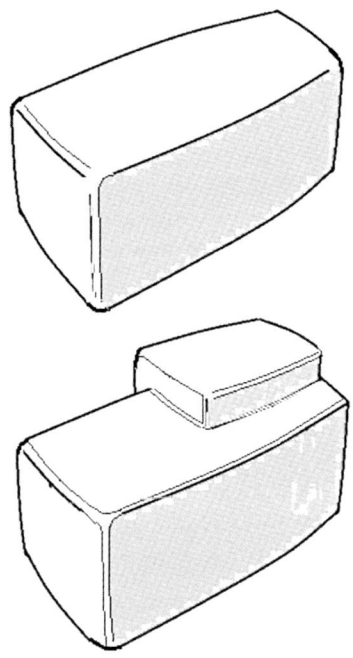

Furthermore, the curvature transitions from one surface to another should be smooth. The same applies to the compound form (Figure 3.25, bottom right), which in reality will have much more complex main volumes than are shown. The section lines A to B and C to D (and other intersections not marked) will need fillets where they intersect. These fillets should flow smoothly from one to the other. Also, the radiused edges or fillets will need to flow smoothly into each other where they meet up, such as where the volumes are joined (Figure 3.26). The ambition is smoothly changing curvature from one surface to the next such that the reflections will show smooth curves, too (as in the car in Figure 3.19).

The basic problem looks like this (Figure 3.27).

This awareness of curvature and smoothness of the surfaces is needed to ensure that the three-dimensional version of the drawing looks as refined as possible and that the shapes are well integrated and homogenous-looking. In terms of acceleration, the change in curvature from the main sections of the curves is handled so that the transition between them is gradual and even.

A fairly refined example of an essentially monovolume form would be the telephone shown in Chapter 1 (here again, as Figure 3.28, for convenience).

The handset has the same kind of radiused edges as the body of the phone. The main surfaces blend neatly and consistently into the sides.

Many items of consumer electronics could be taken as an elaborated example of the smooth integration of a large number of subsidiary forms. Bearing in mind a) the complexity of the main volumes and b) the requirement for curved

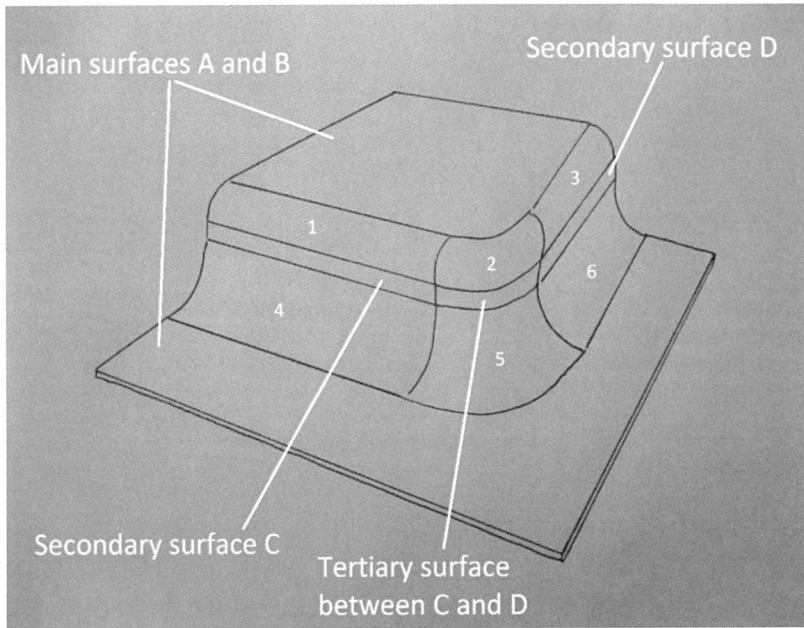

3.26
Main surfaces A
and B, secondary
surfaces C and D
and their associated
fillets 1–6. Notice
that the fillets 1–6
are smaller than the
radius of the tertiary
surface between
C and D. The fillets
1–6 dominate the
secondary surfaces
C and D and the
tertiary surface
between them.

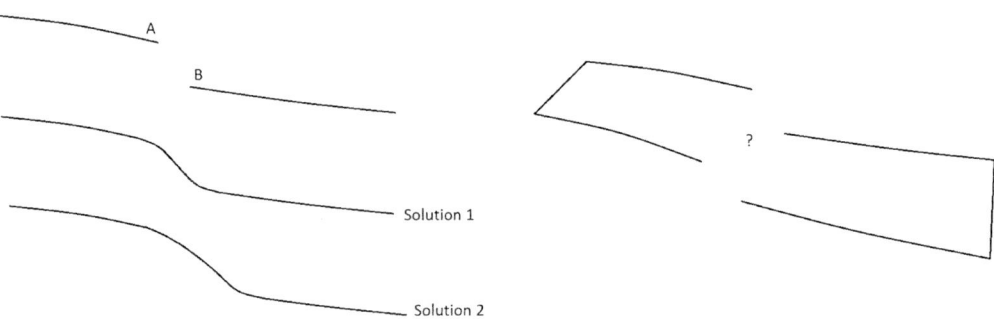

3.27
A pair of curves with a gap (left) and two ways of blending them below (top left). Either is possible but both should be smoothly blended to the main curves. The same curves in 3D (right). Joining the two sides is about finding a smooth blending surface.

edges and surfaces, the designer has two missions. Mission one with respect to the flow of surfaces is to avoid unsettling highlights which distract from the main message of the form. Mission two, somewhat harder to achieve, is the condition where the highlights and interplay of shadows are, in and of themselves, pleasing to consider. It is this surface quality that makes Ian Pollock's *Waiting Swallow* (Figure 3.20) so successful, along with the Georg Jensen jug (Figure 3.21). These two missions can only be completed reliably when the main surfaces and blends (fillets, radii, etc.) can be made to flow together in an organised and coherent way.

3.28
The phone again.
Two monovolumes
from Chapter 1, the
body of the phone
and the handset.

3.7 FILLETS AND MAIN SURFACES: RELATIVE PROPORTIONS OF FILLET TO THE MAIN SURFACE

For any set of primary and secondary surfaces, there are usually a few options about how to manage the fillets which form the rounded edges that result from their intersection. A lot of the time, a designer will not be actively thinking about the fillet radii. Instead, they may be intuitively sketching the overall shape and detailing the rounded edges and corners by eye. But there are times when what looks good on paper will not turn out so well in three dimensions. This section offers an overview of the underlying range of choices in how to treat the rounded edges and corners involved in compound forms.

Figure 3.29 shows two main surfaces, A and B, and two minor or secondary surfaces C and D. They have associated fillets, blending from the upper surface A to the lower surface B; there is also a fillet blending from C to D, going around the corner, a tertiary surface. In Figure 3.29 the overall effect is one of soft forms because the radii of the fillets are quite large. The same underlying surface geometry could also be blended using small fillets all around or just smaller fillets joining A to B (see Figure 3.30).

There are intermediary possibilities, too. Figure 3.31 shows some main choices. Notice that the combination of a large main curve or surface plus big fillets leads to a shape where the distinction between main surface and fillet is hard to see.

The matrix of possible combinations of fillets for a simple cube is 180 based on three fillet sizes (small, medium and large). The number of options rises to 240 if one includes the option of no fillet but a sharp edge. Not all of them

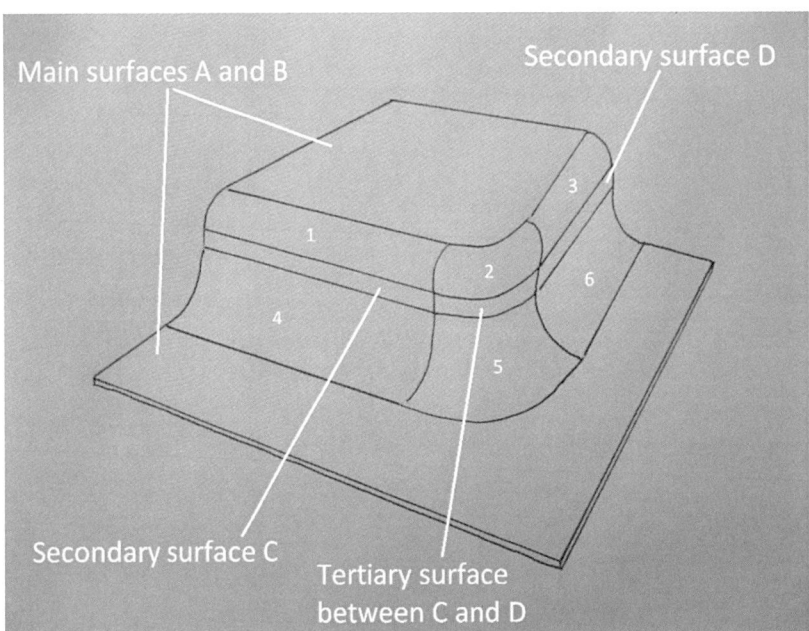

3.29
Fillet combinations
on a rectangular
form.

Main surfaces A and B

Secondary surface D

Secondary surface C

Tertiary surface
between C and D

1 2 3 4 5 6

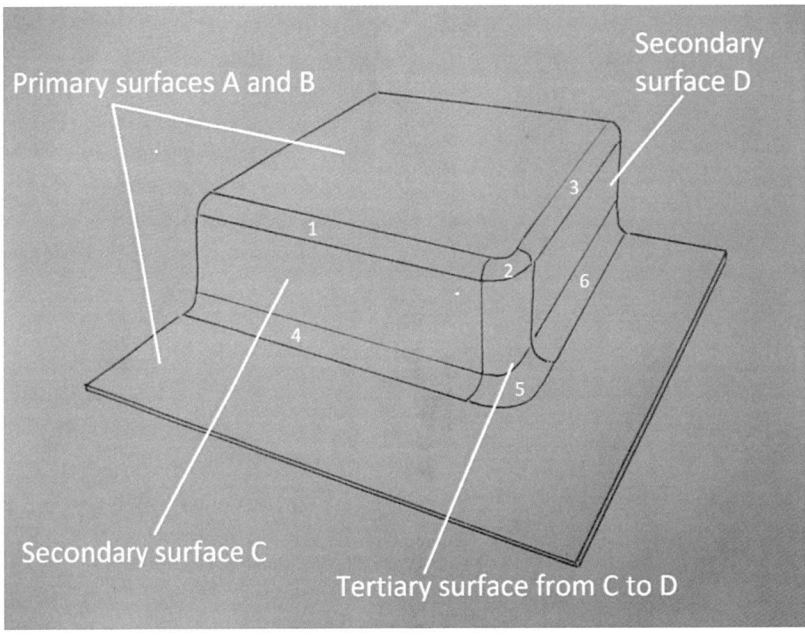

3.30
Smaller fillets but
the same underly-
ing geometry as in
Figure 3.29.

Primary surfaces A and B

Secondary
surface D

Secondary surface C

Tertiary surface from C to D

1 2 3 4 5 6

3.31
Combinations of fillet sizes and curves/sections of surfaces. A section through two flat main surfaces with a small fillet (top left). Two main flat surfaces and a very dominant, large fillet (top right). How large fillets on very curved surfaces produces a very inflated-looking shape (bottom right).

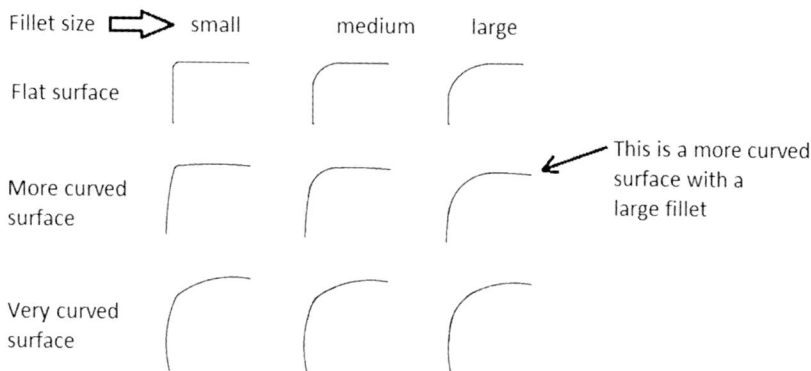

will be aesthetically satisfactory, though. For an object composed of several interesting volumes, the number of combinations rises to a very large number indeed. What this means is that there is likely to be a lot of work done to find the optimum combination of fillets to finish a form made up of several volumes.

From this we can understand that 'form language' consists (in part) of different combinations of curvature characters plus fillet sizes. We can use Table 3.1 to analyse existing styles in terms of the main surface and fillets. A style is, in part, a set of consistently applied rules applied to the relations between surfaces. The fact that there is an infinite number of possible fillet sizes between none and relatively big means the designer has a fair degree of latitude in styling an object, even if this were the only parameter. That said, most people don't make very fine distinctions, so most possible combinations will be judged as 'fairly angular', 'fairly rounded' or somewhere in the middle.

Notice how only the main extremes of combination are immediately distinguishable: totally angular, middling or so rounded as to be perceived as a quite organic shape.

EXERCISE 3.8

The aim of this exercise is to show how the rounded edges of an object are resolved.

Take a photo of a product with rounded edges.

Using that photo as an underlay, draw out the main surfaces on a sheet of tracing paper.

The result should be an angular outline like the marked-up image below.

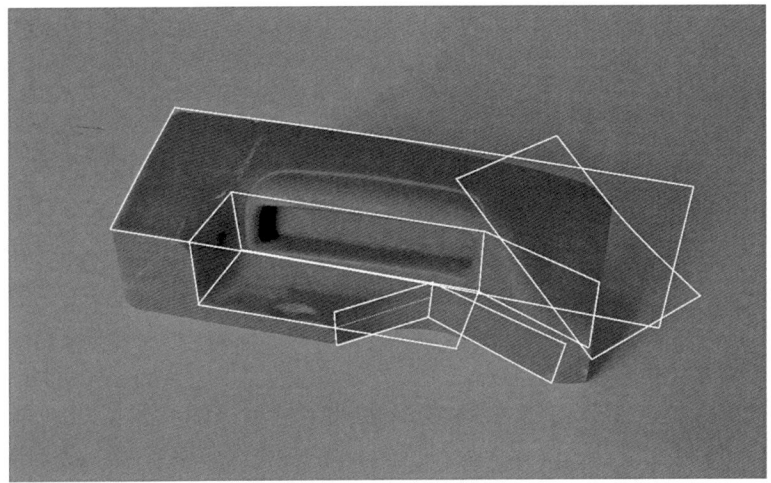

Redraw the shape using one fillet size so they are all smaller. If possible, try doing one version with much bigger fillets and one with very small fillets. What happens to the shape? Are there problem areas where you find you can't draw the fillet as they intersect?

You might notice that the main surfaces and fillets are 'balanced' on the original object such that the fillets are not independent of the main surfaces' character but seem harmonious: about the right size for the amount of curvature on the main surfaces.

3.8 SHADOW: POSITIVE AND NEGATIVE SPACE

This section refers back to the work of Cheryl Akner-Koler discussed in Chapter 1. In *Three-Dimensional Visual Analysis* (1994) she presented the notion of positive and negative space. Positive spaces are those which expand out into the viewer's line of sight (Figure 3.32, left). They are enclosed by a surface. Negative spaces recede – think of a cavity, or the space inside a soup bowl (Figure 3.32, right). Positive and negative space are relative. Seen up close, parts of a car body or casing of a product like a hairdryer are negative spaces in that they recede in relation to the surrounding surfaces (e.g., a recess or scallop feature). As a whole, and seen from a middle or greater distance, the object defines a positive space in that it advances towards the viewer or expands out from the object's centre. Buildings can be considered as positive spaces; their facades are where positive and negative are played off against each other. This section considers the interplay of positive and negative space (see Figure 3.33).

3.32
A positive space (a volume) on the left; a negative space (inside the bowl) on the right.

3.33
A form constructed
of three main
surfaces. The light
source is located
above the object.
Notice how the
shadow falls on each
panel.

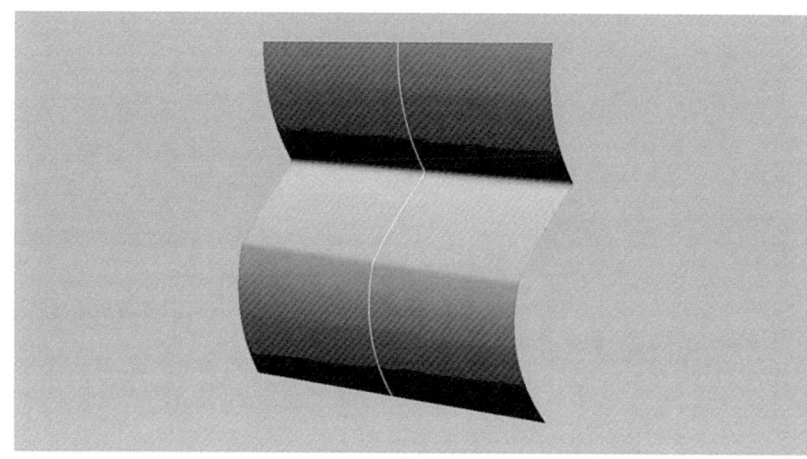

Designers who wish to work with light and shade draw the boundaries between bright and dark with lines before applying marker or its equivalent in a digital drawing tool (see Figure 3.31).

For this drawing (Figure 3.34), the entire car is a positive form in relation to its empty surroundings. Where the wheel arch lips meet the main body is a negative space in that it is locally concave. Another negative space exists at the base of the door, positioned to accentuate the sense of forward motion and directionality. The height and depth of this space matter, as does the surface quality of the transitions.

Figure 3.35 shows a clay modeller working on a 2011 Ford Focus. Notice the relation of light and dark. There are particularly dark areas on the zag feature under the side mirror (A) and at the bottom of the front door (B). The one at the

3.34
Drawing (2018) by
Lucian Bové, Renault
Advanced Design
Centre.

3.35
Side profile of clay model of 2011 Ford Focus. (Image courtesy of Ford, Europe)

base of the door (B) fades gradually from front to back. This would be shown in surface analysis as a very slow decrease in curvature as the feature fades out as it heads rearwards. The effect of the shadow is accentuated by having an upwardly facing surface (C) below it.

The revised image (Figure 3.36) shows more clearly the proportion of light and dark on the body-side.

Car designers can experiment with the proportions of shadow and light by putting black tape on clay models. (This method of working is not well

3.36
Side profile of clay model of 2011 Ford Focus, with shadows marked out.

documented, so I have had to alter the photo to suggest the effect.) The designer would use wide black crepe tape where the darker areas should be. The tape guides the clay modeller in the process of adding or removing material to achieve the desired surface contour. This is of concern to the designer because how the surface is divided into lighter and darker areas affects how the length and height proportions of the car are perceived. Figure 3.37 shows the general idea. The upper and lower oblongs have the same dimensions. The lower one is divided into lighter and darker regions, and this gives the impression it is slightly longer (left to right dimension).

3.37
The lower oblong is perceived as being longer than the upper one, more so when seen in isolation. In fashion, stripes on shirts are usually vertical, to make the wearer look a bit thinner.

The object in Figure 3.37 is two-dimensional, without depth. On a three-dimensional object such as the cylinder with rounded ends (Figure 3.38), the effect of varying light and dark would be the result of the reflection of the grey ground and white surroundings.

3.38
A rough sketch of how a cylindrical form with rounded ends reflects a grey area below it.

On a more complex item the surfaces will reflect the available light and objects in their surroundings. Usually that means the surfaces facing the sky will be brighter. Those surfaces reflecting the darker horizon will be darker. Figure 3.39 shows the same object reflecting panels above and below it.

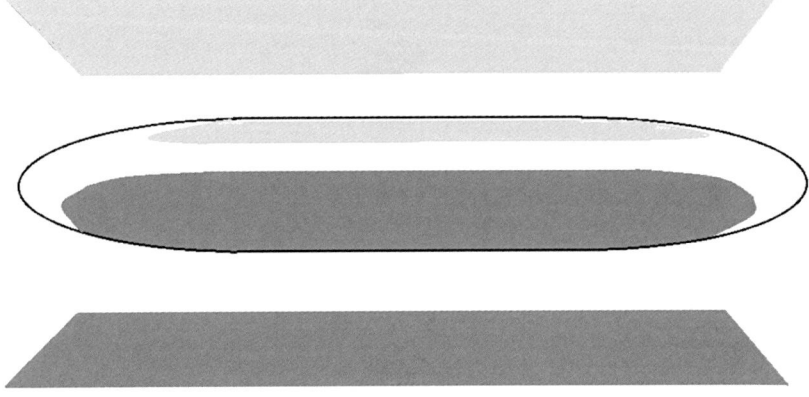

3.39
A rough sketch of how a cylindrical form with rounded ends reflects a grey area below it and a lighter one above it. Look closely at an object near you and see where the light and shadow fall.

The lightness (or reflectivity) of the ground will affect the degree to which the downward-facing areas will be light or dark. Figure 3.40 shows the interplay of light and shade on this clay model by Ford, Europe.

The essence of the foregoing is that an object must be considered in relation to the potential light sources in its environment and that horizontally aligned surface changes (e.g., convexity and feature lines at or near horizontal) can help convey impressions of length and directionality. Vertical lines will accentuate height.

3.40
Clay modelling exploring the relation of light and shade. What is the effect of those bands of light and dark? (Image courtesy of Ford, Europe)

Variations in surface form result in variations in light and shadow. On the complex, three-dimensional example (see Figure 3.36), the boundaries of these darker and lighter areas are, in effect, sections through the three-dimensional form. The curving, bending, flowing highlights on the surfaces constitute the last layer of lines on the object.

Unlike the shut lines and graphical elements, they are very fluid and dynamic. The same object can look quite different depending on the way the paint reflects the light source and the angle from which the object is viewed. The quality and character of those boundary lines matter as much as the fixed lines of the geometry. The quality of those reflections is directly related to the curvature characteristics of the surfaces. The shinier the surface, the more it will reflect and the greater the need for good curvature characteristics.

EXERCISE 3.9

These activities are meant to draw your attention to the boundaries of light and shade as lines and also to the possible drama of sharp contrasts which show the surface sculpting.

1. Do a light and shadow drawing of an object. That means draw the main lines and then show the lighter and darker regions. You can use one shade of light grey for simplicity. Then apply section lines showing the surface variation.
2. Take a late evening walk and examine the play of streetlight on parked cars. How does the different light alter your perception of the cars compared to daylight?

3.9 BRINGING THE IDEAS TOGETHER: A REMINDER OF WHY IT MATTERS

Having introduced the concept of curvature, surface quality and positive and negative space, the final step is to consider again when these concepts come into play. The reasons are threefold: functional, aesthetic and semantic.

First, for functional reasons, many products of industrial design are given radiused edges (see Figure 3.41). This is usual for products made of plastic and metal and things like glass containers. The rounded corner is more robust, more pleasant to touch and easier to produce than a sharp edge. The main surfaces may and usually do have some degree of curvature too. Curvature adds strength as the surface is able to withstand greater loads; the curved form directs forces away from the point of application much as an arch can bear a heavier load than a straight beam. These requirements lead to forms which have some gentle curvature and rounded corners and edges.

Second, after more than a hundred years of industrial product design, rounded forms with fair curvature are customary and have been elaborated beyond merely fulfilling their basic functional purpose. The end result is that whether by eye or with the help of CAD analysis, corners in all their variety are characterised by their smoothly continuous curvature. Industrial design has certain formal expectations, and the functional requirements for radiused edges and curved surfaces is now a style itself. Its apogee is the industrial design corner where multiple surfaces meet smoothly with good flow of light and shade.

Third, there are semantic reasons. With the elaboration of the form comes an understanding of that form as a signifier. As is typical with any kind of functional requirement in design and architecture, the form may be elaborated over time and take on further, semantic meanings. For example, the wooden posts of ancient temples were replaced by carved marble and, after that, the forms

3.41
Digital radio with
some rounded
corners.

underwent further and further development, taking on symbolic meaning that
eventually became detached from the function of holding up a weight. In the
case of cars, a long bonnet was once necessary for a powerful engine. Over
time it became expected that a powerful car had a long bonnet even if engine
technology did not necessitate it. Modern audio technology means a large case
is not needed to hold speakers, but there are still numerous tabletop playback
devices that look as if they have large speakers within their casing. In a similar
way, the rounded edge of the industrial designed object is now a style, with
its original purpose mostly forgotten. The rounded edges of industrial design
products are not only functional but also have come to mean or *imply* functionality. The digital radio in Figure 3.41 has notable rounded corners seen
from the front view, but the designer has also chosen to omit them on the
edge of the front face, making a statement about the form; earlier attempts
at the industrial design form language would have probably included radiused
edges wherever possible to carry the message clearly: that this is a modern
plastic product and respects the requirements of the material. VW's car, the
Up! of 2011 (Figure 3.42), eschews the strongly swept forms typical of automotive design and uses the restrained semantics of classical industrial design

to convey coolness and utility. This car is styled as if it is a very high-quality appliance.

3.42
2011 VW Up! – Apparently simple forms which are really very refined indeed.

The chamfer under the car's side window is gently concave to force the shadowing and make the contrast with the body side seem sharper. The designer has ensured that the chamfer will very often have a different reflective intensity than the main panel. This is achieved through angling the surface and by making it a negative or hollow surface (see Figure 3.43). The main surface of the door will seem brighter than the chamfer under most light conditions. The door panel and adjacent surfaces have enough curvature on them to create subtle surface richness as well. Between the two wheels the door panel curves outwards both vertically and horizontally (by a very small amount). While seemingly simple, the Up! is a highly finessed piece of design.

The main point to remember from this section is that smooth surfaces with good transitions make for a harmonious interplay of light and shade. The edges of lighted and shaded areas are themselves curves which will be judged as fair or not. Light and shade add a dynamic element to the perception of the form. They need to support the message of the product (is it cool, kinetic, friendly, robust and so on). Ripples and sudden changes of form are distracting. But beyond that, one is interested in the feeling of the curves (consider the ones shown in Figure 1.47 in Chapter 1).

3.43
Sketch section of chamfer feature on the Up! car (Figure 3.42). It's the way the curvature is handled on the chamfer that makes the reflections behave as they do. The hollow/convex/negative form of the surface and its upward-facing angle ensure it catches the light under most conditions.

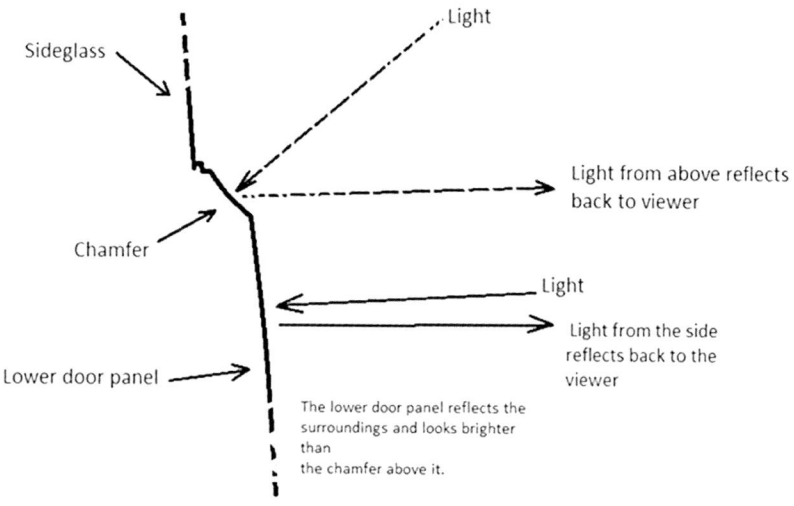

Sideglass

Light

Light from above reflects back to viewer

Chamfer

Light

Lower door panel

Light from the side reflects back to the viewer

The lower door panel reflects the surroundings and looks brighter than the chamfer above it.

EXERCISE 3.10

The aim of this exercise is to become aware of light and shade and the drama possible by their use and to see how a surface's quality is revealed by the use of sharp lighting to make sections on the surface.

Find an object that you wish to study. Obtain a strong light source, for example, a bright desk lamp. Photograph or just observe the object as lighting goes from direct to oblique. How much of the design's identity remains when there is almost no light falling on it? Examine the surfaces up close for smoothness and irregularity. The diagram below shows the object in plan view and two positions of the lamp (direct and side lighting). You could use this method to study an object you have modelled. Alternatively, you can rotate the object in relation to a bright light source, for example, strong sunlight.

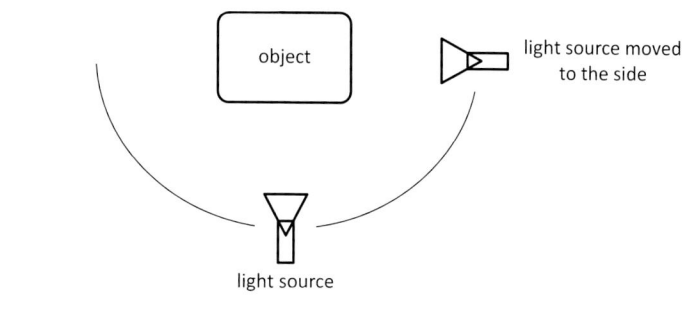

object

light source moved to the side

light source

3.10 JOINT LINES

Like applied graphics (see Section 3.11), joint lines may be among the first features sketched by a designer. They are a notable element of most plastic products and are usually obvious elements of many other finished products. However, they are dependent on the primary surfaces they are derived from. A wobbly surface will yield strange joint lines and it is likely that good, clean surfaces will make it easier to design nice-looking joint lines.

Joint lines are where different components meet at assembly. On a drawing, the joint line is usually represented by one single dark line which describes the shadowing created by the gap between two parts. What the joint line is in reality is the meeting of the edges of two or more parts. The curvature of joint lines needs to be considered as carefully as the curvature of the surfaces and the visible edges. Although small in proportion to the overall size, joint lines can have a profound effect on the way a design is perceived. The image in Figure 3.44 shows an Electrolux vacuum cleaner with the joint lines and with them removed. The joints are probably gaps of 1.5 mm or less but have quite a large effect on the definition of the form. The version without the joint lines seems less defined and less satisfactory.

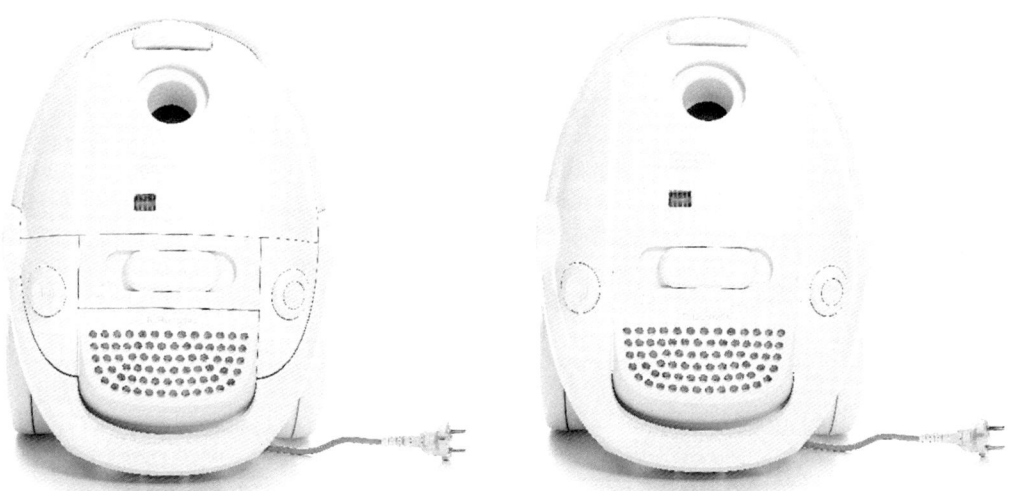

3.44
Electrolux UltraSilencer Pia Wallen edition, with joint lines (left) and without joint lines (right). Are the dots form or are they graphics? (Image: Electrolux)

Joint lines, shut lines or panel gaps are dependent on the geometry of the surfaces they bound. Good, well-structured main volumes will afford the construction of joint lines that themselves will have good curvature characteristics. It is safe to say that if the joint lines are well resolved they would look satisfactory if considered in isolation. Problems may occur with joint lines where the curvature of the joint line (say, seen from a certain main view) is not an accord with the surface it is located on. Usually a successful panel gap is constructed by the intersection of a plane (a construction plane) and a curved surface (the object being designed; e.g., Figure 3.45).

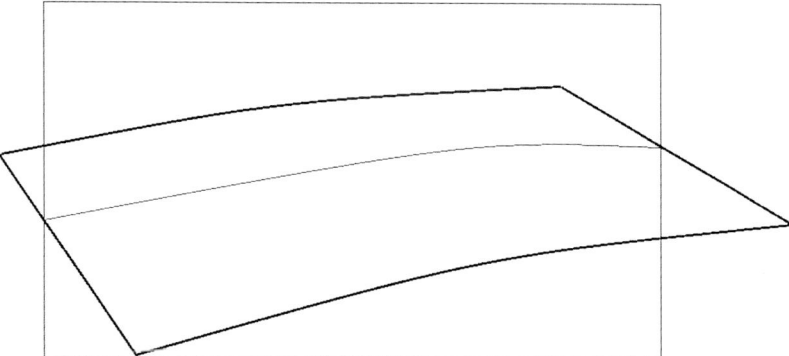

3.45
Sketch of a plane intersecting a curved surface.

Less successful is where a curved construction surface (say, an irregular curve) intersects with another curved surface (the object being designed). From certain views the resultant intersection may seem to have inconsistent curvature; that is, it may look oddly flat or have an unwelcome change in curvature seen from some angles (see Figure 3.46).

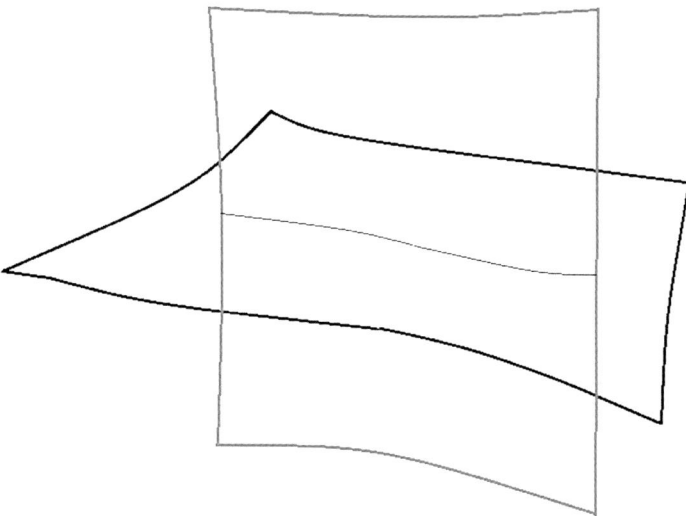

3.46
Two irregular surfaces intersecting. The curve resulting from this is also uneven and will have changes in curvature that look disturbing. Is it supposed to be a line or a simple curve? It is ambiguous. You would not draw a curve like that. This kind of thing crops up when you try to turn a sketch into a 3D model and the intersection of two forms do not (in CAD) yield a good intersecting curve. Another, more typical hazard is where two surfaces intersect to make a curve that is a bit flat somewhere along its length.

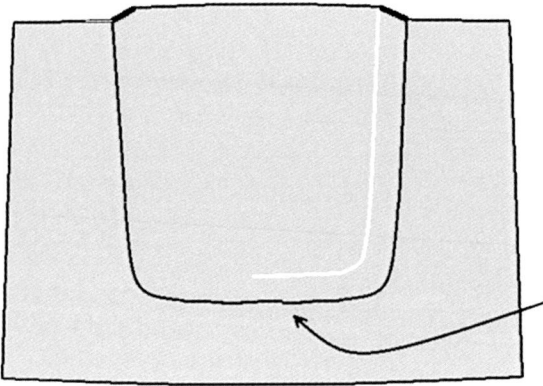

3.47
Intersection of one or
possibly two uneven
surfaces.

One of the surfaces
is not curving steadily
here which results in uneven
acceleration or uneven change
of curvature where the
surfaces meet each other

Examples of wobbly lines are rather hard to find in production as they seldom make it past the design stage. What one does see is marginal examples where the curve is unsettling if not mathematically 'wrong' and the wrongness might not be immediately apparent (see Figure 3.47).

EXERCISE 3.11

The aim of this task is to investigate the role of joint lines on a three-dimensional form. This exercise shows some of the puzzles involved in creating good surface lines on a large and complex form. This exercise requires a CAD programme.

Make some shapes such as those below. The one on the left is a surface curved in one direction only.

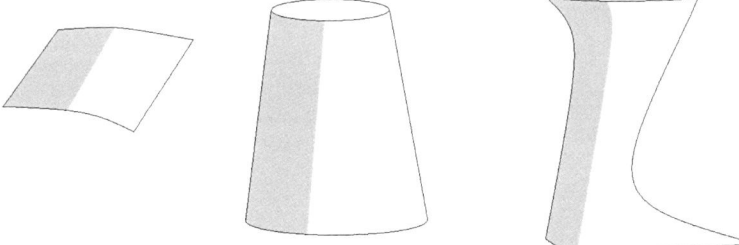

Create a plane and use it to inspect the type of curve created when the plane intersects with the test objects. What you should be looking for here is the fairness of the intersection. The general principle is that two regular surfaces will usually create an intersection which has smooth curvature and no 'wobbles'. A plane can intersect with a free-form surface to make an intersection which has good curvature, but it depends on the angle of incidence. Two irregular surfaces will seldom meet with a regular intersection. It can happen but mostly won't.

When you have explored the different ways in which the intersection planes and surfaces interact with the model shapes, use those intersections to make curves on the surface. You can use the curvature analysis tool to check the quality of the curves.

For the sake of completeness, I will note that that joint lines and materials or colour differences in aggregate make up what are called 'graphics'. That is a shorthand term for the collection of lines and distinctly different materials or colours used on a product.

This conception of graphics works as if there is an underlying, massive three-dimensional form over which are superimposed secondary visual elements: the materials, surface finish and edge lines of the individual components.

While graphics are an important part of the final form of an object – and, indeed, many sketches start out with the graphics as the dominant element – graphics are dependent on and subsidiary to the three-dimensional geometry of the object and then the edge lines that divide it into different parts. For the moment, I will assume that attention paid to three-dimensional form and the edge lines of parts will be sufficient to account for the acceptability of the graphics. Put another way, if you design with graphics in mind, the surfaces they are situated on must be good. If you design good surfaces, the higher the chance any resultant graphic effects will be good.

3.11 APPLIED GRAPHICS OR DECALS

The diagram at the start of this chapter (Figure 3.1) shows that graphics are something of an outlier. Graphics (as in coloured films or painted areas) are appended at the end of a chain of several other interconnected aspects of curvature. In one sense this is ironic since applied graphics might be the first thing one notices and might be the first element sketched. Yet without having good surfaces to be applied to, applied graphics risk being inadequate or visually disturbing. Hence, this topic has found itself as one of the last items dealt with in this chapter.[4]

Applied lines and other graphics are secondary to the underlying form they are placed on. Applied lines include stripes, graphics and lettering (Figure 3.48). When used, they need to be used with some care. The two principal concerns are a) the placement in relation to the surfaces they will occupy and b) their relation to the overall form.

As in the previous sections, curved surfaces can be thought of as having principal axes in relation to which the surfaces fold (see Figure 3.41). Printed text is typically aligned horizontally and read left to right. In the case of text placed on a product's outer surface, that general principle needs to be balanced with the demands created by the geometry of the underlying surface. In the

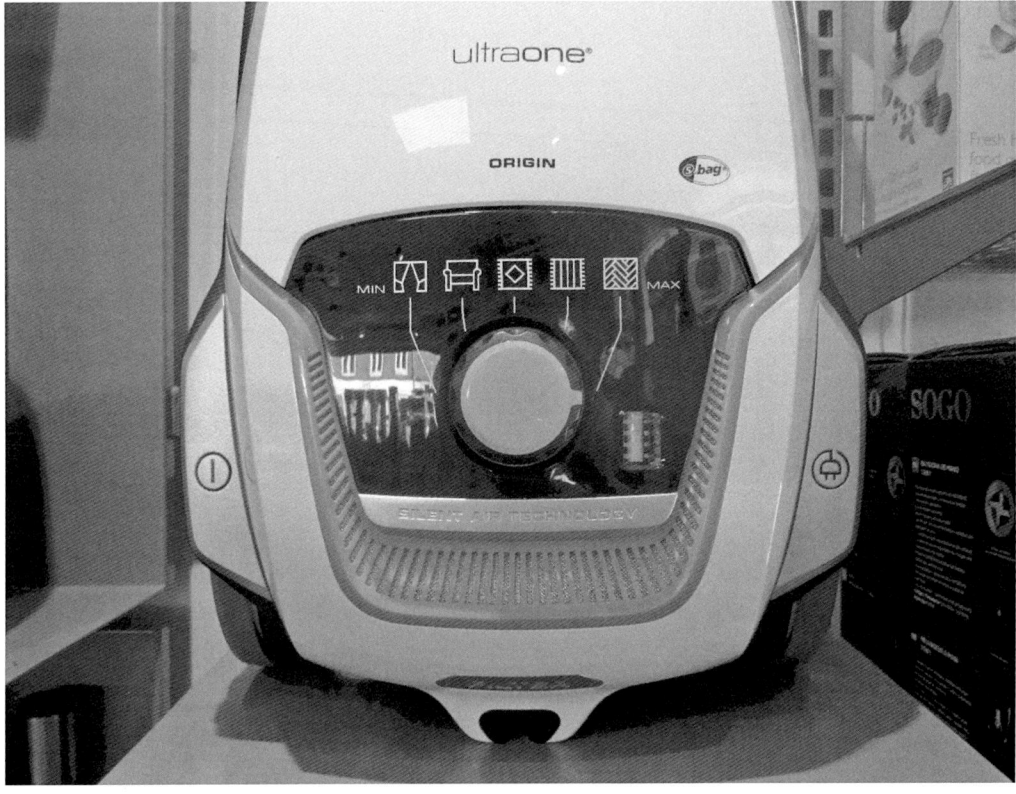

3.48
Applied graphics on a vacuum cleaner. The coloured elements can also be considered as graphic elements. Notice also the way the part joints correspond to the coloured areas. Colour areas that cross part joint lines need to be painted coatings or applied films and are costly to do and prone to abrasion during use.

case of very sculpted surfaces, horizontal placement may not appear horizontal owing to the optical effect caused by the surface curvature.

SUMMARY

Just as a musician is interested in notes, a designer should be interested in the technicalities of curvature.

Being able to master curvature puts the designer in a position to first remove unwanted distractions but also to attain some level of what is called beauty. Kant (1928) developed the concept of the aesthetic moment which we may understand as a 'wow!' experience. When 'wow' happens, all considerations of utility and meaning are set aside due to the force of the visual experience. There is a very strong correlation between the surface refinement and the aesthetic impact perceived by the viewer. At the very least, an understanding of curvature will help the designer avoid unwanted surface effects so that the attention of the viewer is not distracted from some other positive aspect of the design.

NOTES

1 This parable owes something to the concept of useless work expounded by Pye (1968).
2 CAD construction is often based around planes governing the main views (profile, top, elevation) and planes which control the principal features of the object. It is also affected by the way fillets or rounds are applied to edges and corners.
3 If you use parchment tracing paper and flexible black crepe tape of 1.1 to 1.25 mm thickness, you can do a very smooth curve through the points. Chartpak makes flexible black tape which is also good for making full-size line drawings. It is halfway between drawing (it is two-dimensional) and modelling (it allows great precision).
4 In the lecture upon which this book was based, applied graphics and split lines were explained first and briefly before I moved on to technical matters of lines and surfaces.

FURTHER READING

Westermann, J., Gardner, P., Sutherland, E., White, T., Jordan, K., Watts, D., & Wells, S. (2012) Product design: Preference for rounded versus angular design elements. *Psychology and Marketing*, 29(8), 595–605.

4 Craftsmanship

4.0 INTRODUCTION

This chapter presents examples of specific aesthetic problems that relate to the craftsmanship of designed objects. Craftsmanship isn't always something you can draw, yet it's often the quality of craftsmanship that affects people's verdicts on an object, whether a building, a shirt or a home appliance. Some aspects of it are to do with how much control the production engineers can impose on the manufacturing process; for example, a part junction that looked good on paper and as a model might turn out to be hard to consistently make well during mass production. Other aspects are the details that designers directly control, such as how well part joints gel with the rest of the design. It's often very small-scale things (below 5 mm in size) that are hard to capture on paper, which is why a special form of critical imagination is needed to catch craftsmanship failures and also to spot opportunities for improvements that can set your product apart. The tricky thing is that you have to see what is not there and imagine something that does not exist and may not do so until the first prototypes are made.

Design researcher Christopher Frayling (2011) listed ten different interpretations of the word 'craftsmanship', mostly to do with things being made by hand and involving expertise of some form. I take the word to relate to the quality of production and assembly, both of which might be done by machines in large volumes. An injection moulded children's toy might demonstrate high levels of craftsmanship. Playmobil and Lego spring to mind. One of the lovely aspects of these toys is the remarkably high standard to which they are produced despite the very large numbers made. Craftsmanship does not only have to do with something made manually, one item at a time. This chapter is about details to keep in mind when designing objects to look as well made as they can be – attention to small but important elements that signal care and effort in the design process.

The emphasis in this section is more on the empirical than the theoretical and, as such, it may seem to resemble a random collection of subjects. While it is very satisfying to be able to firmly bolt insights into some form of intellectual framework, some topics seem not to belong to a very large, well-defined

DOI: 10.4324/9781003183303-4

general class. This chapter's unifying principle is variety. This is particularly true of the articulation of edges and joints (discussed below), which itself is a theme made up of heterogeneous examples.

If there can be one main outcome of this chapter's selection of design problems, it is to understand that considerable effort can be expended on tidying or 'editing' forms. This is so that they make the most sense and that unwanted ambiguity is minimised. If the design problems described here are hard to classify, they compensate for this by cropping up often enough to be worth addressing.

Figure 4.1 shows one way the elements of the chapter can be related to each other. The underlying theme is how to divide a surface into two or more parts that fit can be fitted together and look acceptable. It is about attention to fine details in the context of the whole product.

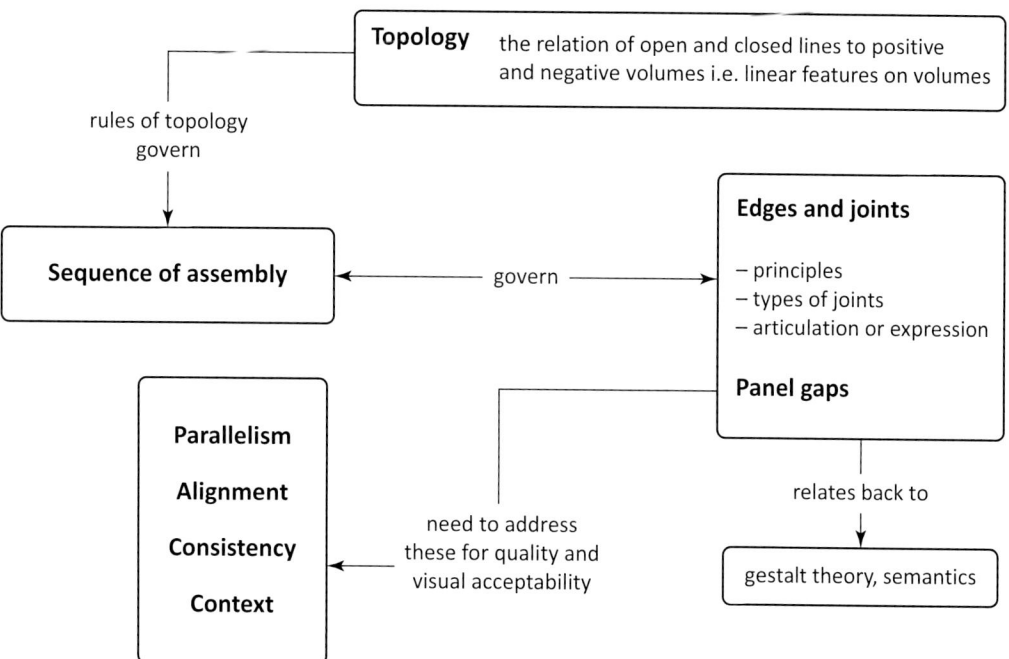

4.1
How the elements of the chapter relate to each other. Semantics is dealt with in the next chapter.

Perhaps the simplest point to be taken from this section is that it is best to design with craftsmanship in mind to avoid unpleasant surprises that come too late to be fixed easily. To do that it is best to take design for assembly into consideration as soon as possible. That means deciding the principal sections for the joints where the object's components meet. This makes clear the construction principle and its general effect on the local appearance of the part.

The details in this chapter are almost unavoidably tedious to the layperson and many designers. However, mastery of the art of the industrial design joint (together with the radius and fillet) is one thing that distinguishes the master designer from the novice. Joints and junctions have semantic as well as functional significance: done well, they signify care and quality. Done badly, they ruin an otherwise good design.

The aim of this section is *not* to be dogmatic nor to claim any of the general rules of thumb are unbreakable laws. Any or all of them could for various reasons be ignored. However, knowing what those general guidelines are in the first place makes it more likely that the flouting of them will work as intended.

4.1 JOINTS AND JUNCTIONS

This category is simple enough to define. It is about designing things to look acceptable when the parts are put together (see Figure 4.2). Under that rubric comes the immense diversity of possibilities of the combination of geometry and the huge number of materials involved. Further complicating the matter is that edges and joints are situated in the context of the object in question. They relate to the semantic aspects of the design as well as basic manufacturability. It is not possible to consider the rightness of the joint in isolation from the function, geometry and material of the things being joined.

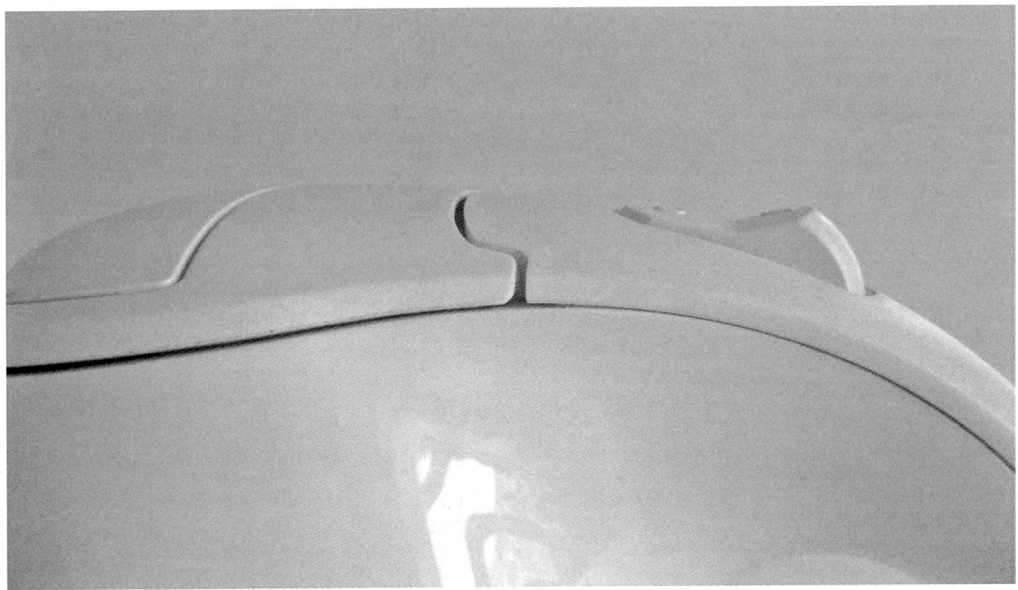

4.2
The landscape of joints and junctions on an electric kettle, as seen from 5-cm distance.

Joints have a great number of functions. ISO 3447 defines them as having … capacity to withstand compression, bending, shear, torsion, vibration, impact, abrasion and creep, to support joined components and resist differential deformation of joined components; to have a specific minimum life and permit, dismantling, re-assembly and replacement.

(Sebestyen, 2003, p. 63)

That is a lot to ask, and it can be hard to anticipate all problems, but some are easier to spot than others, especially with experience.

Frampton (1995)[1] presented the matter of how parts meet, using the term *tectonic*:

Tectonic becomes the art of joinings: 'Art' here is to be understood as encompassing *tekne* and therefore indicates tectonic as assemblage not only of building parts but objects … as soon as an aesthetic perspective - and not a goal of utility – is defined that specifies the work and production of tekton, then the analysis consigns the term to aesthetic judgment.

(p. 4)

Frampton went on to explain that form and material and technique makes a triangle with tectonic in the middle. Tectonic is about how the materials are put together. Tectonic unites the material, the form and mode of construction (Frampton, 1995; see Figure 4.3).

Frampton was concerned with the way the appearance of the object relates to structure. In architecture this would be exemplified by the use of a distinct material for load-bearing elements and another for cladding or non-structural elements.

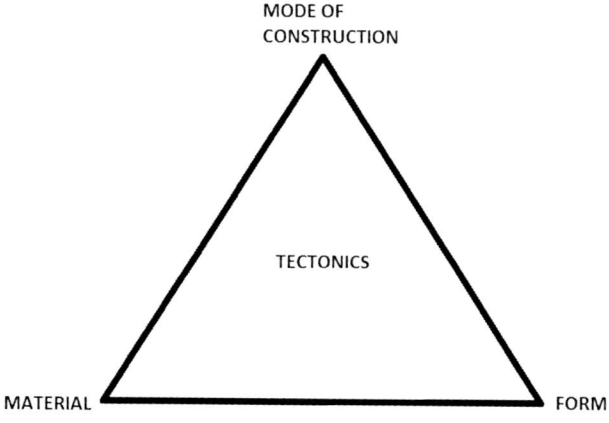

4.3
The tectonic triangle, after Frampton (1995).

EXERCISE 4.1

The aim of this task is to make you aware of how things are put together affects the appearance (and vice versa). While this section places a lot of emphasis on the kinds of joints used in casings (architectural facades and the outer surfaces of consumer appliances), joints of a dazzling range in variety exist in other areas of product design: every combination of everything.

With a camera, go and document all of the types of joints, junctions and juxtapositions you can find in your surroundings. How do buildings meet the ground? What kind of legs do chairs have and how are they attached? Luggage is made of other types of hard and soft materials. How are these materials combined? Which joints around you are fixed and which are movable, such as hinges and pivots? Are the objects joined equals or is one subordinate (e.g., a door handle is subordinate to a door)? Are the joints under stress and, if so, in what directions are the stresses? How are the joints articulated, if at all? What is the difference between the way fashion designers use stitching versus architects and designers?

This is a very open-ended exercise, and the visual data can be analysed in a wide variety of ways. Don't be surprised if you don't reach a definitive point where all of the types of joints are documented and classified.

 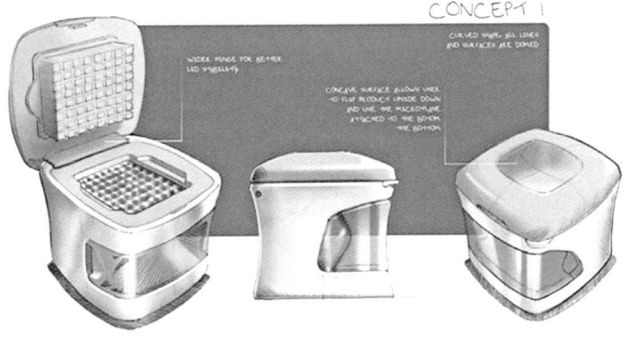

4.4
A building seen as a whole (left) and a product seen as a whole (right). Arguably, the nature of the joints in the building is not as critical to the overall perception of the form as it is in a product. (Building image: Wiki Commons. Product design sketch courtesy of Kitchen Innovations and Spark Innovations, Canada)

In the case of architecture it is more likely to be the material which is doing the work of expressing the idea that each part has a distinct function: concrete lintels, brick walls, steel pillars (see Figure 4.4 and see if you agree with this point). Prior to modernism, there was more room for giving individual parts expression; for example, sculpted lintels and well-elaborated doorways. In the case of product design, what Frampton (1995) calls 'tectonics' are related to functions that are sometimes structural but are more about utility, broadly defined. Due to the scale of buildings, the impact of the lines making up the joints is not as great as it is in product design. It's bluntly a matter of scale. From the distance at which you can make out a whole building, the precise nature of the joints diminishes in importance. If you are up close, you get a weak sense of the whole building, and the crudity or otherwise of the joints in architecture is also overlooked for the most part since we don't expect to touch the building (see Figure 4.5). In contrast, at the scale of a product we sense the overall form and the fine nature of the joints at the same time. We read the object as if we expect to touch it, and we probably will.

4.5
Detail of Lighthouse apartment complex, Aarhus, Denmark. Many small-scale components dominate at human scale, particularly the joints. Is this what was shown in the planning application and advertising material?

The principal aim regarding the resolution of the interaction of joints and junctions is to make them look consistent with the gross form of the object. And it is to ensure the lineaments of the joints do not interfere or confuse the way the overall form is perceived. Ideally, the form of the joints will be pleasing in themselves. I dealt with the curvature quality of such lines in the previous chapter. In this treatment, I am looking at other aspects of the geometry of the lines and the main forms. This matters from an aesthetic perspective and from the point of view of perceived quality:

> The way parts fit together is affected by geometric variation, for example gap, flush and non-parallelism of split lines. If a relation deviates [too much] from its nominal value, this affects the visual quality appearance, defined as: the … impression of quality by, just observing the product.
>
> (Wickman and Söderberg, 2003)

The main drivers of how joints are resolved are as follows:

1. the order of assembly of the parts
2. the material
3. production method (e.g., stamping, injection moulding, welding; if stamped or moulded, the draft angle)
4. the function of the part and the whole object
5. aesthetic demands
6. design for disassembly/maintenance
7. cost.

The sequence of this list is arbitrary; any one factor can be given priority, but all must eventually be accounted for. In a resolved design, all of these factors will have been balanced. Some may be fairly tightly determined in advance and revised only when other factors force changes. For example, even if low cost is given priority it might be discovered that the lowest cost requires too much aesthetic or functional compromise. The design is then reworked so it is cheaper but also acceptable.

Whatever the priorities may have been during the design process, in a finished design it may not be at all obvious what compromises have been made. The result ought to seem like a seamless fusion of elements. One might be able to determine by deconstructing a product how it *could* have been made, but this casts no light on how plans for appearance or production might have changed during the design process.

EXERCISE 4.2

This exercise is intended to show how the appearance of an object is closely related to how it is put together.

Obtain a product of moderate complexity; for example, a hairdryer or item of furniture. Try to establish the order in which it was assembled. Make a sketch of the main components including the assembly direction. Then alter the assembly sequence and assess the impact this would have on appearance. Is it even possible to alter the order of assembly?

Case: A student of mine designed a coffee percolator with each visible part to be assigned one of three colours. Together we analysed how the parts were to be assembled. Assuming that the heater core and base were the first to be put together, we discovered a problem with the proposed arrangement of the colours. The design had to be reworked in order to avoid odd-looking details along the lines where parts met. The elimination of unwanted 1-mm details required a thorough and fussy revision of the design.

Find a household appliance like a kettle or even consider a car interior (the dashboard, for example) and see if assigning one part a new colour creates odd or unsightly junctions.

With other considerations operating equally, the principles for joint and junctions are summarised as follows:

1. As small a gap as is possible between parts.
2. Constant gap condition is preferable to variation.
3. Parallelism is desirable (see Figures 4.38 and 4.39).
4. Neat alignment of parts.
5. Avoid intersection of joint lines at oblique angles (see Figure 4.29).

This list, if satisfied, would be enough to ensure any plain, tidily engineered product would be acceptable. In addition, junctions may need to be articulated or given some expression to lift the design above the level of banality. How might we go about avoiding nasty surprises emerging at the end of the design process? The use of exploded view diagrams is one way of planning for assembly and visualising the details.

The exploded view diagram (see Figure 4.6) sits at an uncomfortable inter-section between the work of the aesthetically aware designer and the more technically driven work of the engineer. This type of drawing often occurs towards the end of a long process involving user studies, theme sketching and functional prototyping. In a sense it is somewhat unavoidable that this is the case, since so many factors have to be considered prior to being able to tightly define the overall appearance of the object. However, the exploded view diagram also forces the designer, perhaps for the first time, to deal with issues affecting how the object is subjectively perceived, how it is seen in the first instance. It is thus both the end of a design process and also a foundational statement of the object's final intended form.

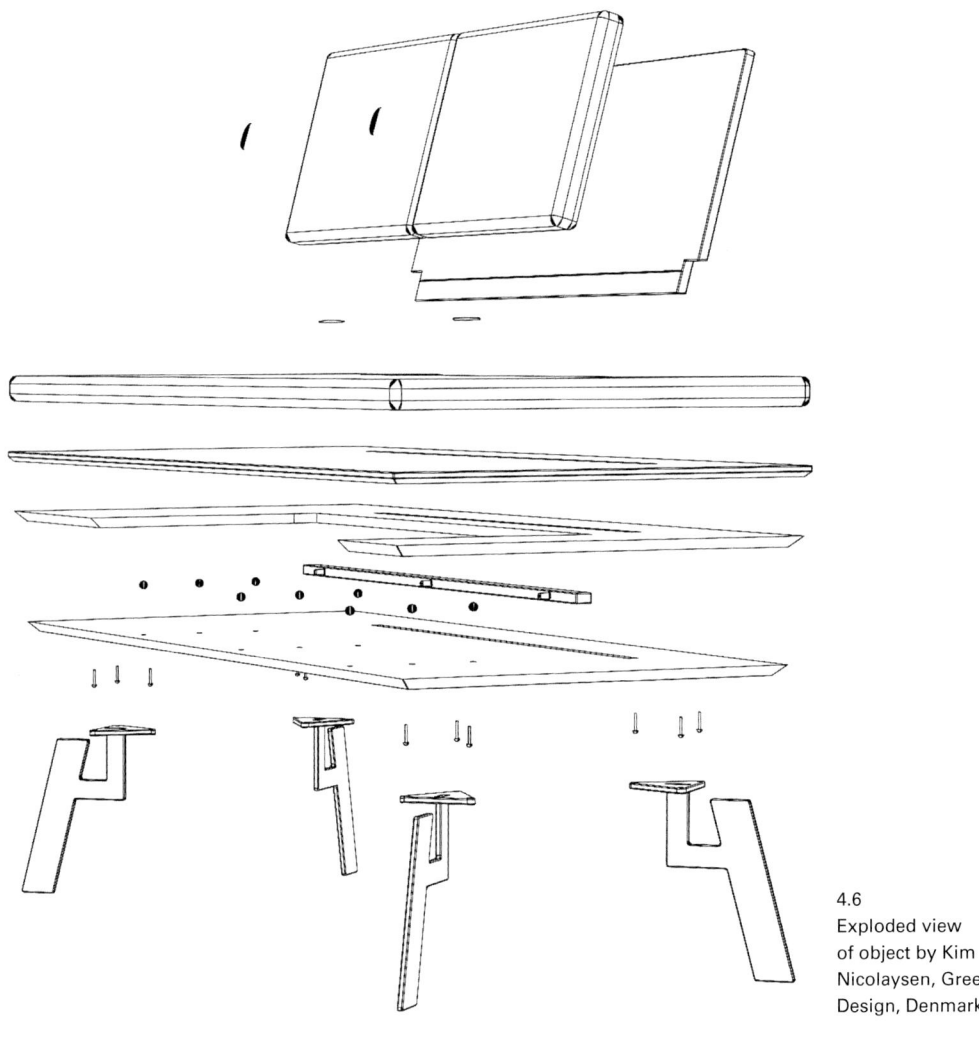

4.6
Exploded view
of object by Kim
Nicolaysen, Green
Design, Denmark.

The exploded view diagram can be used in conjunction with principal sections to show how a product is put together. Together they give a very clear idea of what a joint will look like. The exploded view shows all of the parts and the principal section zooms in on the way they fit together.

This sketch in Figure 4.7 shows a panel with two-colour inset. The principal section A–B shows that the part numbered 2 snaps into place in the part numbered 1. The coloured detail (numbered 3) is bonded to 2 on the inside. This tells us about the assembly order: bond part 3 to part 2 and snap part 2 into part 1. Maybe this sequence is inefficient. If it has to be re-ordered, the design may look a bit different. The alternative principal section A–B suggests that if glued in place, part 3 will not be flush with the surrounding surface 2. It's attention to this kind of tedious detail that can lift a finished product from mediocre to good or from good to delightful.

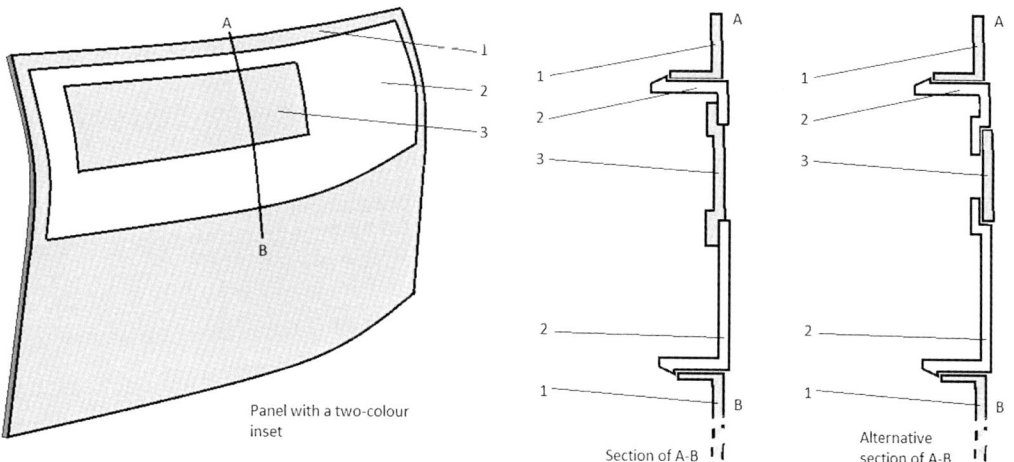

4.7
A sketch of a three-part item (left) and two alternative principal sections (middle and right) that illustrate how the item might be assembled. Making sketches like this requires you to confront the details that can help or hinder the way the user perceives the object. Notice particularly how part 3 relates to part 2.

There exists an interesting and unresolvable question concerning joints and the absence of joints. On the one hand, there is a visual economy in using fewer parts so that the user is not distracted from the overall form and so the production costs can be kept lower than otherwise. On the other hand, even were it possible to have an object devoid of visible joints, it may take on a rather unconvincing, toy-like or massive quality.

Two examples of the massive quality of jointless construction could be considered a) the Pantone chair (Figure 4.8) and b) the bumper of an inexpensive car (Figure 4.9). Many people have an intuitive sense of how much effort was expended on a design. One way of putting it is that information (detail) implies quality or visual interest. The visual interest of Pantone's chair from 1965 is the magical lack of joints made possible by the then-revolutionary method of manufacture. However, in the intervening years, injection moulded chairs have become ubiquitous and are not now viewed as quality items. Is Pantone's chair still visually acceptable? The tentative answer is yes, because the sculptural quality is still appealing.

In the case of the economy car's bumper, the designers have gone beyond the amount of ribbing needed to ensure rigidity and have included fake grooves that suggest a higher part count. A car from a more expensive brand actually *has* a higher part count, even if the economy car shows this is not needed from a functional point of view.

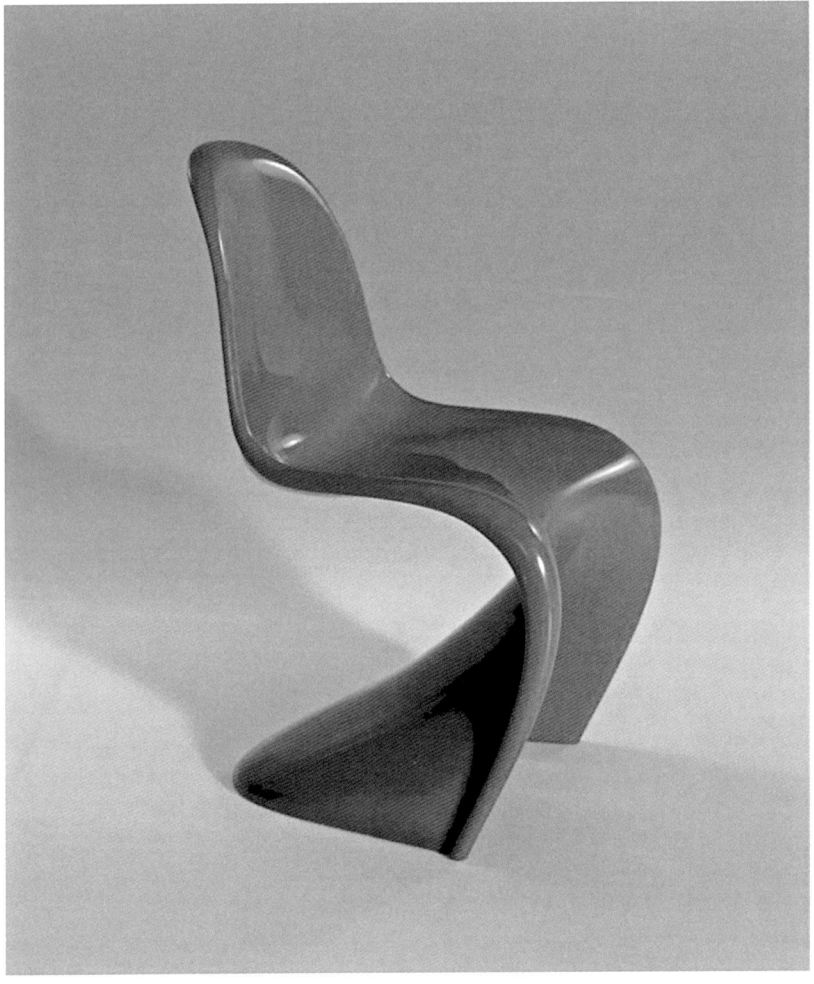

4.8
Pantone chair
(1965). (Image: Wiki
Commons)

4.9
The front bumper of an inexpensive car. Note the grooves under the head-lamps which suggest separate parts even though it is a one-piece moulding.

EXERCISE 4.3

The aim of this activity is to compare the main form of an object with and without joint lines.

An interesting experiment is to do some purely sculptural studies of ordinary consumer objects. Produce some sketches of an object (one you draw yourself) but do not put any part or joint lines on the sketches. When you have produced a shape that is pleasing, do another version but prepare it with joint lines showing where the component parts could meet. You need to imagine roughly how it would be put together on a production line. Something small like a pocket camera has a lot of lines. How do they affect the perception of the object?

Notice the difference in the character of the two sets of sketches. Referring to Figure 4.7, try making some principal sections of how the parts fit together.

Having looked at some of the general principles of joints and junctions, let us now consider how two plastic parts might be joined, using the simplest possible junction geometry.

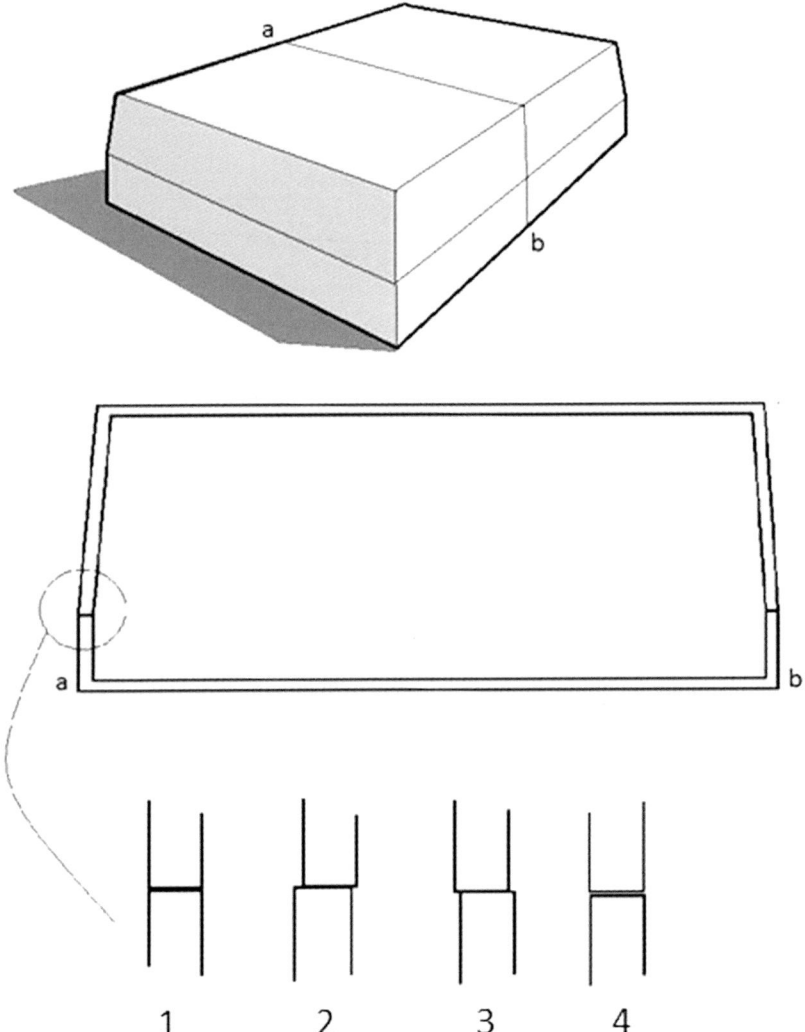

4.10
A sketch of 'clam-
shell' type of box
(top). Notice the
section a-b (middle).
At the bottom is a
section through the
meeting of the upper
and lower parts. It
shows failure types
2, 3 and 4. In number
1 the joint is suc-
cessful: a flush fit. In
the others, the parts
don't meet success-
fully, and 2 and 3 are
offset horizontally.
In 4 the top doesn't
land on the bottom
part so there is a
gap.

Figure 4.10 shows a section through a basic clamshell closure. The upper part mates to the lower part along a plane. A sketch of the product would initially show a line where the two parts meet. But engineering meets aesthetics in this case because the joint has to be resolved in such a way that the two parts can be fitted reliably and quickly. The basic joint, as shown, can fail in at least three ways. Possible failure stems from the simple fact that mass-produced parts are seldom 100% identical. Depending on the quality of the plastic and how it cools, the parts will vary by some amount. Pressed metal can also vary due to spring-back of the metal after forming in the tool. That means the part does not quite retain the shape it has been given when compressed in the tool.

In 2 and 3 in Figure 4.7 the two parts are not meeting flush – there is a small step. One part projects relative to the other. In 4 the two parts don't meet properly in the vertical dimension, and this leaves a gap, probably an uneven one. Failures 2 and 3 can be combined with 4 so that there is an uneven gap *and* the parts are not flush. What this has to do with aesthetics is that an alternative solution is needed and must be highly failure proof and look acceptable. The seemingly harmless line where two parts meet actually requires consideration so that at the very least it does not detract from the overall form.

A possible solution is shown in Figure 4.11.

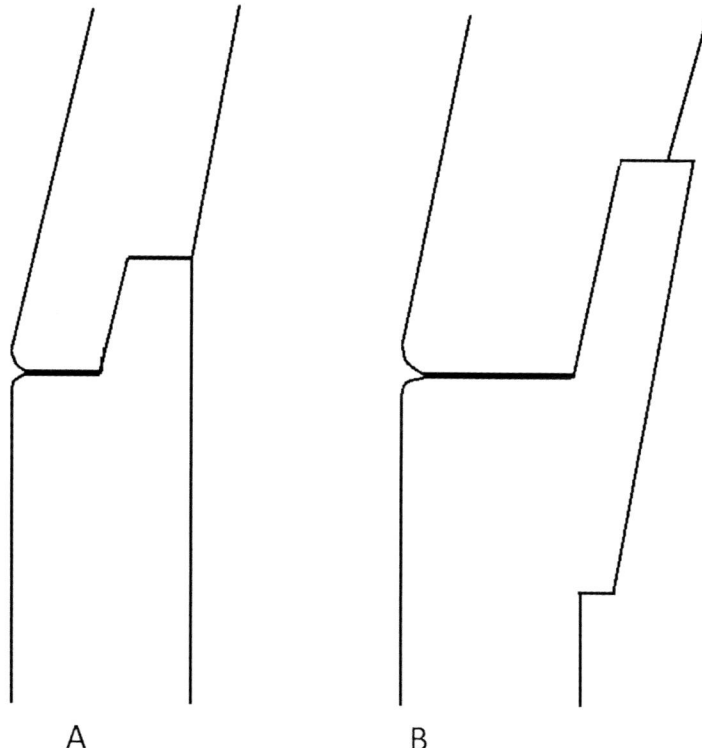

4.11
Basic joint solution.

A   B

In Figure 4.11, version A (left) shows small, rounded edges on the parts so that any remaining uncontrolled gap is harder to perceive. More important, there is a stepped mating surface to help the two parts fit in the desired way. In version B (right) there are radii plus a stepped landing surface and additional material behind the joint line to make the junction more rigid. This kind of detailing is akin to the tailoring that differentiates a good suit from one that is made to a lower standard.

What we get from this is a sense of the nitty-gritty attention to detail that has to be designed for. Even if the lead designer hands over the execution of

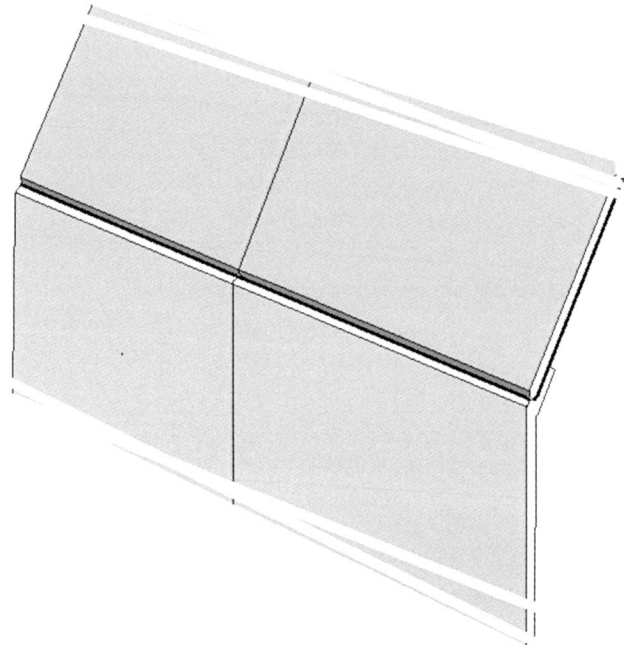

4.12
An approximate
sketch of the joint
concept in Figure
4.11 with small fillets
and stepped landing
surface.

the details to a studio engineer, this issue has to be addressed and checked so that the main design theme is not compromised by inadequate execution.

When designing, one must have some outline idea of how the object is to be assembled, maintained and, with sustainability in mind, disassembled. Not unconceivably, the pursuit of surfaces free of screws and clipping points may look attractive (important for the buyer) but may also confound maintenance and repair because hidden fixings may involve adhesives or clips. Adhesives and clips can make for difficult disassembly and re-assembly. The adhesive joints may fail over time or may be hard to separate without damage; clips might wear after repeated removal and remounting or may simply break upon disassembly. Those factors then affect the product's utility (the part falls off) or pleasantness (loosened clips may cause rattling or squeaking) or durability (the part fails too soon).

Having introduced the idea of design for reliable, robust and aesthetically acceptable joints, we turn to a closer look at some of the solutions available.

The butt joint (Figure 4.13 and Figure 4.16) and its cousin the T-joint could be said to be the most basic types of joint, where one element is adjacent to the other, held in place by pins, glue or gravity. These are characteristic of standard joinery and architecture where there are large, static weights involved. It is a matter of practicality to cut the part to fit and allow a combination of nails or mortar and gravity to keep everything in place.

Of interest to us is the use of this kind of joint in industrial design. It is most often seen in furniture and transport interiors such as the window frame trim

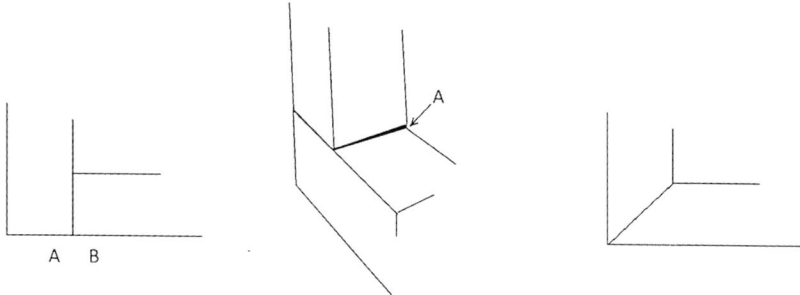

4.13

Butt joint (left). The alignment of A and B might not be adequate; in the diagram (right) the vertical is not aligned to the horizontal. What would a solution to this be? One answer is to have a raised lip around the area where the vertical part meets the horizontal. That would make variations in fit harder to see. On the right, the mitre joint, which requires more precision. The extra difficulty adds to its appeal.

4.14

Train window pillar with butt joint; general context.

4.15
Train window pillar.
Notice the uneven
gap.

4.16
Butt joint (left) on the corner of a storage unit. It is not flush and it is not aligned. The curvature is not that consistent either. A mitre joint on the window frame of a 1982 Volvo 760 (right) (designed by Jans Wilsgaard).

in the Danish IC4 train (Figures 4.14 and 4.15). From a practical point of view, it works well enough in the train's case, allowing the part to be installed towards the end of construction and presumably permitting easy maintenance of the sun-blind spar that runs behind it. The problem here is that the junction with the horizontal element is uneven and, in aesthetic terms, lacks any expression. We can read this as a cheap solution to a practical problem rather than something demonstrative of industrial design finesse. It is unfortunate that the joint is in plain view, whereas if it had been in a less conspicuous place it would have been a more acceptable compromise.[2]

4.17
Sleeve joint. The
distance A to B can
vary according to
the location of the
mounting points.

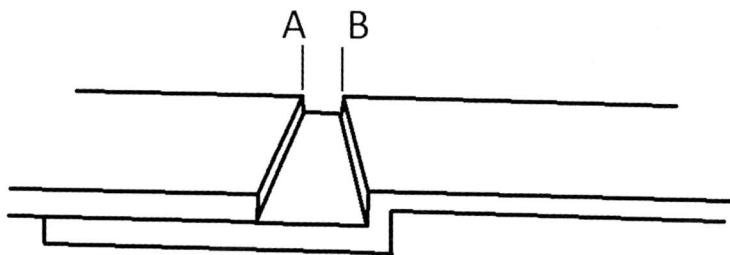

In Figure 4.17 the joint type is a kind sleeve. Gravity and fixings can keep the parts together for a good fit. The version of the sleeve joint in Figure 4.15 can also be used where there is a likelihood of variation in the parts to be united and the substructure. This variation may come from a small difference in the location of the mounting points and in the dimensions of the plastic parts or both.

The shingle joint (Figure 4.18 and Figure 4.19) is a simpler version of the sleeve joint. The principle is that one part overlaps the other. Like the sleeve joint, it allows for considerable latitude in positioning. The disadvantage is that it is nonflush, which is visibly obtrusive, may be uncomfortable to hold, may snag and may trap dirt.

4.18
Shingle joint. The
overlap A to B can
accommodate vari-
ation in the parts'
dimensions and also
the location of the
fixing points.

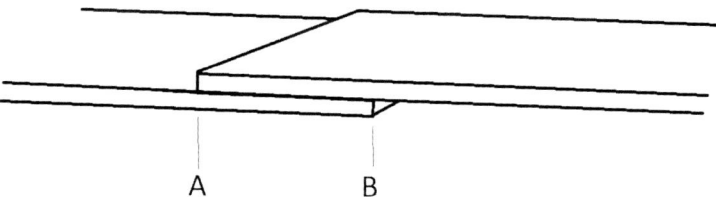

Since we are interested in aesthetics, or how the object is perceived, the shingle joint is vulnerable to criticism for its crudeness. Again, it might be functional, but it is open to interpretation as being a low-cost solution for the problem of joining parts. The intention with joint and junction solutions is that they are as unobtrusive as possible and they make semantic sense. The shingle is open to interpretation as a joint more appropriate to architecture where flushness and human-to-object contact is minimal.

4.2 PANEL GAPS AND SHUT LINES: THE SPACE BETWEEN PARTS

A panel gap is where parts are adjacent and aligned but there is no immediately visible contact between the parts. It is the dark line of the gap that is visible as much as the edges of the parts. These kinds of junctions are typical of applied skins or facades. The shut line is a subset of the panel gap and is found between an opening panel and its surrounding aperture.

4.19
Shingle joint on a
window ledge.

With panel gaps one or more parts are affixed to the substructure either elsewhere (away from the evident joint line) or out of sight, by means of clips or the parts forming a closure (the cover of the water reservoir of a coffee machine, a car door aperture, the cover on a vacuum cleaner housing). Figure 4.20 shows details of a photocopier. Nearly all of the joints visible are shut lines, a necessity given the fact that the machine has so many panels which must be opened for access to jammed paper and errant paper clips. While the empty, dark space of those gaps comprise as little as one percent of the visible exterior of the device, ordering those lines such that they are functional and at least unobtrusive probably consumed a considerable amount of the engineering design of this product.

More tolerant or indifferent observers might say that since this was merely a photocopier, the design of such shut lines is not all that important. Be that as it may, it does illustrate the effect of having a variety of panel gap dimensions, a kind of worst-case scenario. We can also reflect on the meaning of this: a firm more interested in product communication might (and I would say *should*) insist on maintaining standards for managing perceived quality. One interpretation of this landscape of lines of varying thickness is that the producer was indifferent to the customers' perceptions of quality.

Hinged joints involve one part moving around an axis in relation to another part. One consequence of this motion is that it might not be possible to maintain a constant apparent or actual gap condition; in the example in Figure 4.21, the visible gap is wider parallel to the hinge than it is at either end.

4.20
Photocopier exterior panels. Notice the variation in the panel gaps and shut lines.

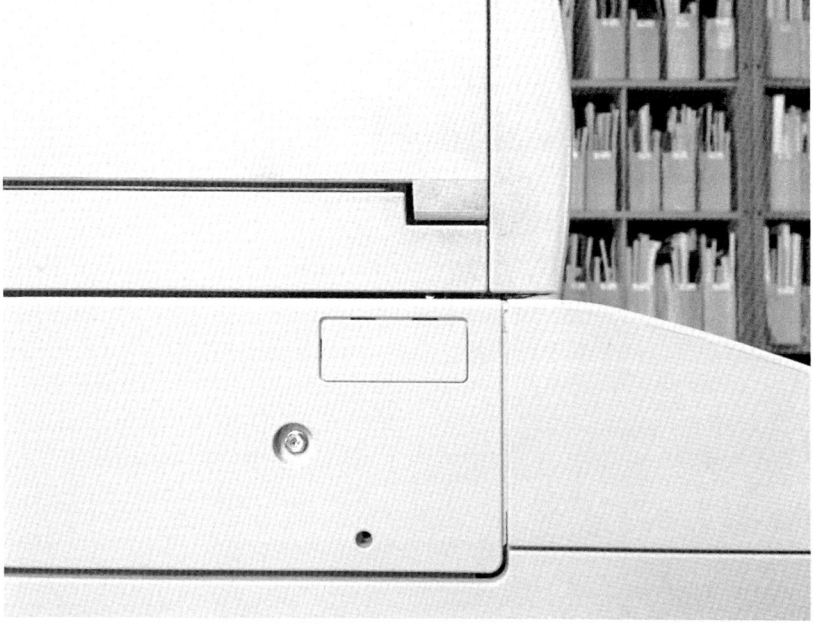

4.21
A handle on a photocopier panel. The gap parallel to the hinge (on the left) is wider than that at the ends. Is there anything else that might be troublesome in the way this handle and its recess has been designed? The rounded corners on the left are sharper than those on the right. Maybe they should have been the same on all four corners. The matter is discussed further in Section 4.3.

Collars are where a cylindrical form, or part of one, is inserted into main surface (see Figure 4.22). The decisions to be made concerning this type of junction relate to:

- whether the joint boundary should be flush or articulated (and how much articulation),
- the size of the radii on the edges, and
- the material choice.

The most understated design would be flush, with minimal radii, and would use matching materials. An example of this kind of joint might be a louvred panel attached to a larger surface. The use of a separate part is most likely to do with the difficulty finding a draft angle suitable for entire assembly or where the louvred panel is made of a different material to the panel it is placed in.

4.22
Collar joint on a food blender. It's the ring-shaped area between the brushed steel panel and the black plastic rotary dial.

In Figure 4.22 the design is likely a result of the designer's wish to articulate the junction of the control with the panel. It permits the use of a second material, a reflective chrome-effect plastic. In addition, the joint is not flush but has relief in relation to the brushed steel it is adjacent to. Figure 4.23 shows a cylinder intersecting with a flat surface with no articulation. It is a very simple solution. It looks as if the cylinder is just resting on a flat surface.

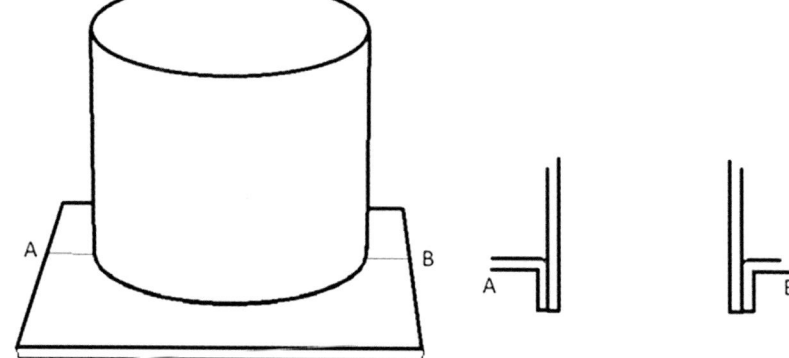

4.23
Intersection of a closed form (here it is a cylinder) and flat surface is in principle the essence of a collar joint.

Figure 4.24 is another way of doing the intersection. There is a raised rim around the aperture which is formed from a separate ring-shaped part (shown in the section on the right). This might be desirable when a different material is needed for robustness or to draw attention to the intersection.

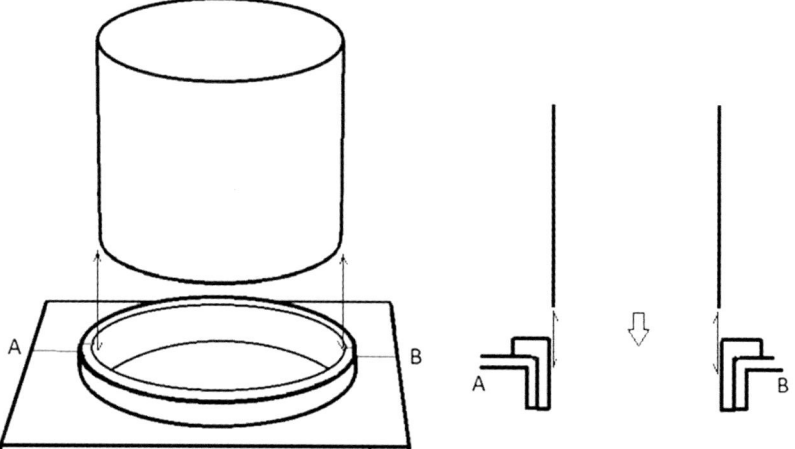

4.24
A simple collar joint. This intersection can be articulated in many ways.

EXERCISE 4.4

The point of this exercise is to show how a joint concept can be expressed in several ways.

The section A–B in Figure 4.24 is not that interesting. Try to sketch other possible ways to articulate this junction. Try 2D sections first (like in Figure 4.23, right side) and then try to draw them from various angles in 3D if that's needed.

As an example, here is a sketch of another version, with a feature to clip the ring into the aperture.

4.3 THE INTERACTION OF JOINTS AND PANEL GAPS – WHEN PARTS MEET

The angle of how joints meet each other matters, especially when considering radiused edges. Consider the arrangement of lines in Figure 4.25. The lines represent panel gaps. A proposed section through the gap is shown, with two fillets of equal radius on either side of the gap.

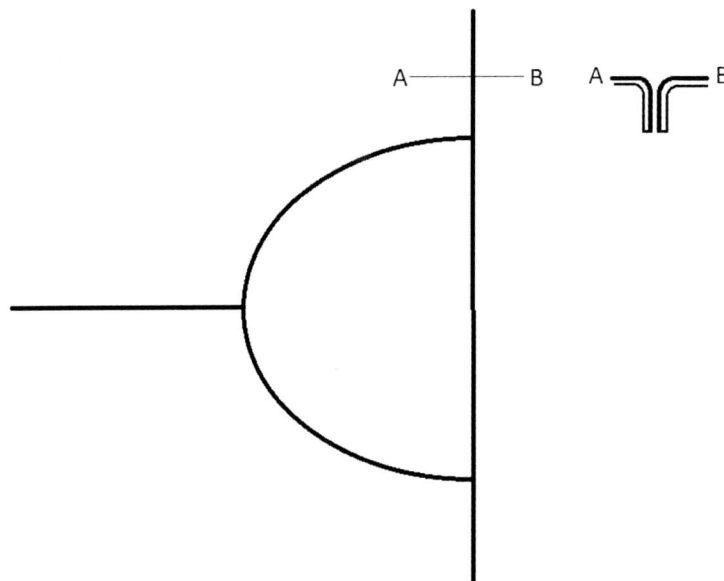

4.25
A schematic layout for panel gaps.

This is how it would look with the panel gap and fillets applied in plan view (Figure 4.26).

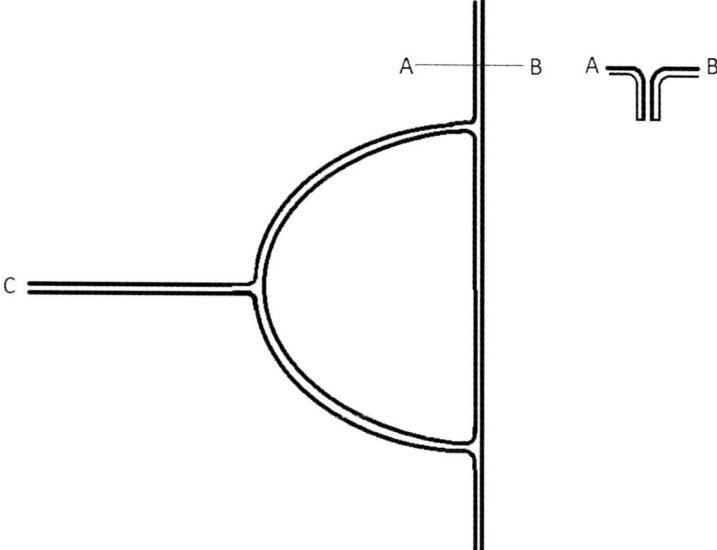

4.26
The lines with fillet and gap applied. Notice the small triangular areas at each junction (1, 2 and 3).

At each junction (numbered 1, 2 and 3) the apparent gap between the parts looks a bit bigger than in the other areas (Figure 4.26). These areas will attract attention, especially as the gap itself will be dark with respect to the sur-rounding surface. Only a smaller panel gap and smaller fillet radii can mitigate this effect. What if a line has to intersect another one at an oblique angle for some reason (Figure 4.27)?

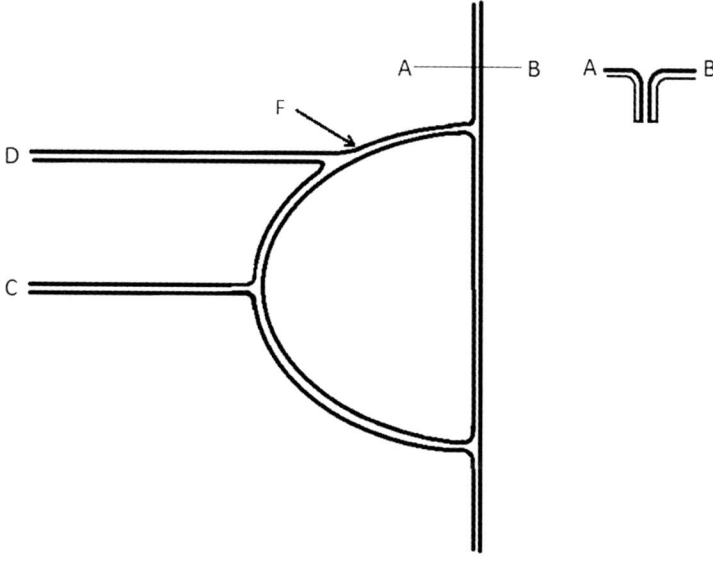

4.27
Where the lines D meet the oval lines there is a quite large triangular gap which is unsightly. The curvature at F also needs attention.

These effects are even clearer in these CAD models (Figure 4.28).

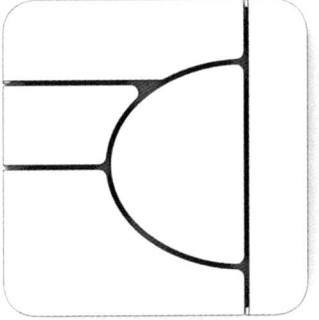

4.28
This is how the schematic lines in Figure 4.27 would look as a 3D model. Notice in the right view that there is a conspicuous gap where the linear gap meets the oval gap. (Model: Patrick Bomholt)

The effect can be seen here on the headlamp in Figure 4.29. The designers may have tried to reduce the apparent size of the gap by eliminating the radius on the upper side of the junction. It does so to some extent but at the cost of looking visually inconsistent. One corner seems not to have a fillet or radius and the other one does.[3]

4.29
The circled area shows a three-way junction between a car headlamp, the bonnet and wing. Do you think the elimination of the radius on the upper side of the junction was a good solution?

One way to reduce the apparent gap at such junctions is to design the lines so that they do not meet at oblique angles. Rather, the lines should best meet at right angles or near to right angles. In Figure 4.30 the lines D are altered to

intersect the oval lines at about ninety degrees (the line g' to h'). The resultant fillets produce a smaller gap (at the cost of introducing a bit more visual
complexity). Figure 4.31 shows how it might look in a 3D model, seen with a
touch of perspective. Figure 4.32 shows a real-world example, in this case a car
bumper-to-wing panel gap.

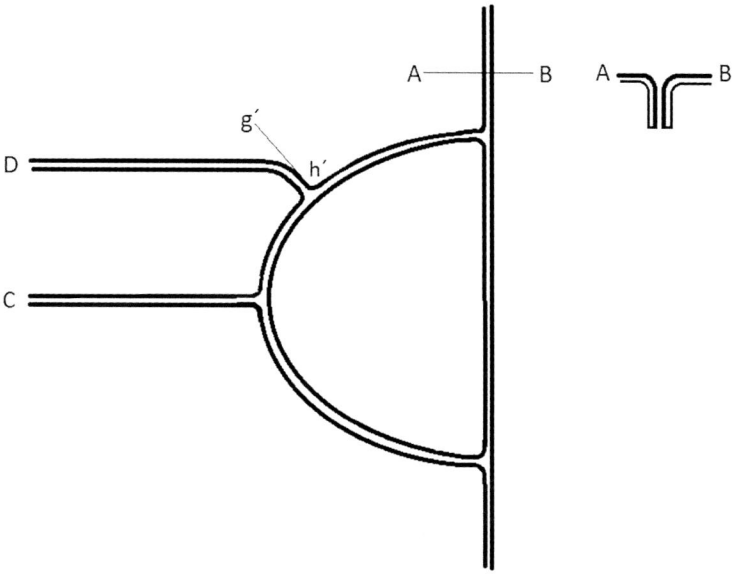

4.30
The lines D meet
the oval shape after
swerving to intersect
at something like
right angles (g'–h').

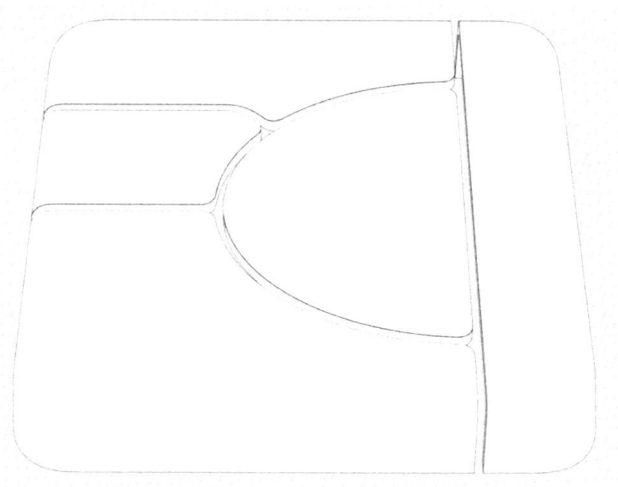

4.31
This is how it would
look at as a 3D
model. Notice how
the triangular junc-
tion is now less con-
spicuous. (Model:
Patrick Bomholt)

4.32
This example is analogous to the schematic drawing in Figure 4.30. The panel gap between the bumper at the front wing intersects the wheel arch at A. This kind of thing is not easily shown in sketches and is often resolved at the modelling stage. The designer needs to be aware of this kind of problem and to anticipate such matters as early as possible during ideation.

The schematic example in Figure 4.26 assumes that the parts have no draft angle on them. If the section A–B had to be open to include the draft angle, the gap would seem even larger. Figure 4.33 shows a section through a panel gap where the edge is angled to meet the requirement of drafting, which is fairly common, if not mandatory, in most production ready designs.

4.33
The gap c–d is the minimum permitted distance between the left side and right side of the gap; due to the edge having to be open for drafting purposes, the gap at the top is wider. The fillets increase this effect by making edges harder to perceive.

What the designer is seeking to avoid here at such oblique intersections are 'blips' which interfere with the perception of the curved lines. The rathole effect, seen at middle distance, tends to look like an unwelcome splotch or blob, for the want of a better term (Figure 4.34).

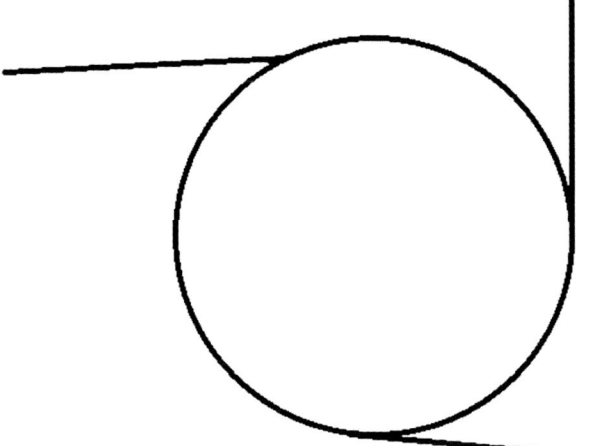

4.34
Lines intersect-
ing with a circle.
Notice the apparent
thickening of the
circle at the points of
intersection ('blips'
or 'splotches' or
'blobs').

As noted earlier in this book, the human eye is sensitive to pointed forms, too. This is possibly because, from experience, pointed forms tend to be asso- ciated with danger. It might also be an empirical effect of noticing the fragility of narrow, pointed forms and their preponderance to catch or snag. Figure 4.34 has three sharp-looking teeth where the straight lines meet the circles.

Finally, the intersection of the lines and the circle interrupts the reading of the curvature of the circle and there is potentially some ambiguity caused by uncertainty over the continuity of the curve. Gestalt theory (see Chapter 2) suggests that a boundary line belongs to the figure; in the case of the circle meeting the lines (Figure 4.34), the boundary line ownership is ambiguous and potentially can be read as shown. Figure 4.35 shows more clearly how the thickness of the circle seems to increase where the lines intersect.

Anthropomorphic theory (see Chapter 1) suggests a preference for even curves, so there might also be an aversion to interrupted or uneven curves, which is what the splotch effect suggests. It seems that the curve is uneven, even if it is not so in actual fact.

So, to conclude this section, joints and junctions are supposed to be robust, visually unobtrusive and designed to be reliably constructed. As stated above, the permutations of factors are too large to allow a general classifi- cation. However, this section ought to heighten the reader's awareness that eventually a given general design theme needs to be reconciled with the ways to put the parts together so that it:

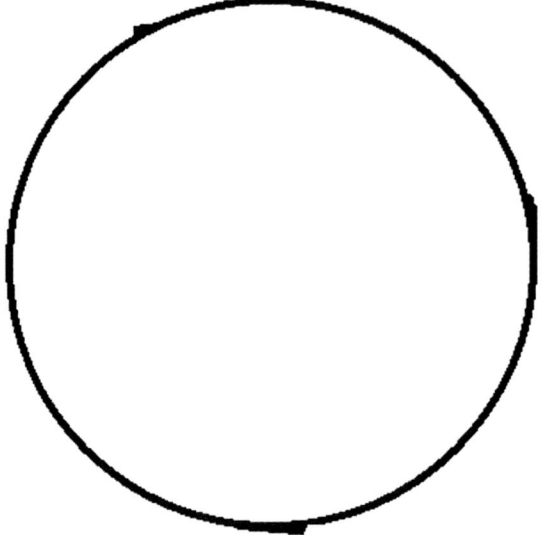

4.35
How the circular part of the form in Figure 4.34 could be interpreted. Following gestalt theory, the outline belongs to the circle. The intersecting lines seem make the outline look uneven.

1. can be reliably implemented,
2. is robust,
3. looks acceptable, and
4. is comfortable to use.

A deep appreciation of this aspect of design has led some car manufacturers to the forefront of customers' estimations of perceived quality and thus to commercial success.

4.4 THE RELATION OF THE GAPS AND JOINTS TO THE ORDER OF ASSEMBLY

The order of assembly might in principle be determined by whether the parts go inside a space as in a vehicle interior or are mounted on a structure. The order of assembly has quite a large bearing on the exact type of joints chosen. These might be sleeve, shingle, butt joints or others. Related to this are the types and conditions of the panel gaps between parts if they are not in direct contact.

The design of an interior of a train, for example, is affected by the fact that the main interior panels have to be installed and then these are followed by subsequent, smaller trim parts. Whilst designers might prefer articulation of joints, consistent gaps and flushness, the demands of assembly could be prioritised for ease of assembly and disassembly.

An interesting contrast in approaches is shown by the interior of the Deutsche Bahn ICE train by Siemens and the Danish regional train by Alstom. The designs were apparently executed to rather different priorities.

4.36
Detail of interior of a
Deutsche Bahn ICE
train. The vertical
part is about 12 cm
across. Notice the
flaring of the trim on
the upper right and
also the visually con-
stant gap condition.

The ICE detail (Figure 4.36) shows complex layering of the parts; where the gap condition could not be assured a slightly larger distance was left between the two parts. It is harder to see if the gap is varying by a small amount. The variation is possible because train chassis are all slightly different, within set tolerances. The variation can be in the order of centimetres overall (not locally). The trim parts do not vary by that much and so the design accommodates these dimensional differences. Notice the slight flaring of the plastic trim at the upper edge, which is articulation. There are no visible screw heads. The photo is taken from a corner on the upper part of the train interior and is not especially conspicuous. A lot of effort has been invested in this small and probably overlooked corner.

The Danish train (Figure 4.37) is designed for easier assembly and disassembly. There are visible screw heads which might be deemed unattractive or overtly industrial. As discussed in an earlier chapter, not all that is functional is acceptable. It is a matter of values, not of fact, though. Designers need to be

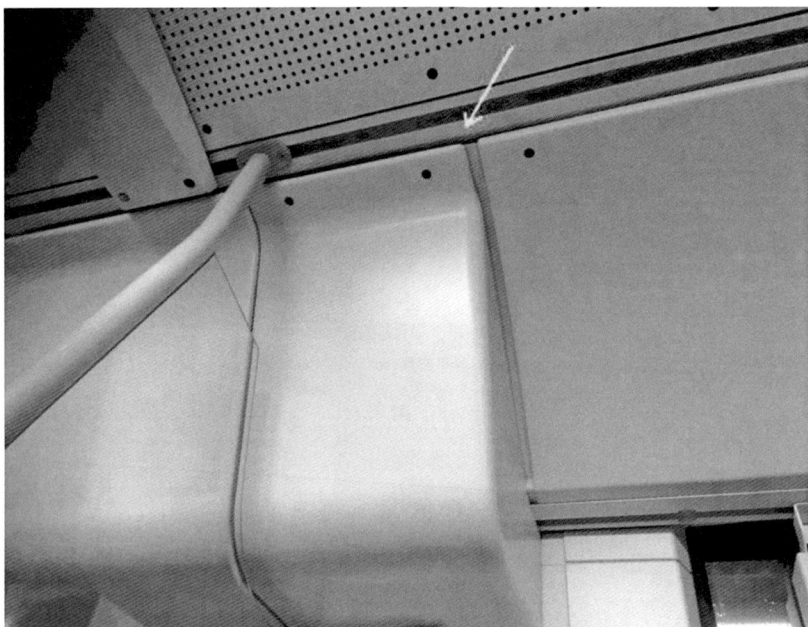

4.37
Interior detail of
Danish regional
train. Notice the vis-
ible screw heads and
sleeve joints.

able to argue for the prioritisation of appearance or functionality, and this can be done with reference to user investigations as well as professional experience.

4.5 PARALLELISM AND CONTROLLED CONVERGENCE

As a default, main feature lines and volumes should be ordered to appear decisively parallel if they are not to be decisively diverging. This rule of thumb must be applied judiciously, of course. This discussion assumes that locally, at least, more order would be desirable. The idea behind this is that the form ought to be clear in its intent and not ambiguous; that is, that the form is clearly parallel and meant to be so. In fine art, ambiguity is often interesting and desirable; in industrial design the product is supposed to convey a clear message of intention and deliberation.

In the example in Figure 4.38 there is a slightly off-parallel box (top) and the adjusted version (bottom). It was possible to adjust the lines so they were parallel. This, as was said, is not always a preferred solution. The alternative to forcing parallelism is to make the convergence more pronounced (Figure 4.39). That eliminates the unsettling ambiguity present in the upper version and would lead the viewer to understand that the convergence was an actively chosen outcome.

The general guidelines to be drawn from the cases of parallelism and controlled convergence are that the design intent should be clear and that unwanted ambiguity is best avoided. There are always exceptions, but they should be examined and considered and not left as default design outcomes. If the shape is not clear, it should be revised to make it so.

4.38
Parallelism. The upper box is slightly off-parallel. The right side and left side are not the same length. This condition is ambiguous and is rectified by adjusting the length so the left and right sides are equal.

4.39
Controlled convergence. The upper box is slightly off-parallel. The right side and left side are not the same length. This condition is ambiguous and is rectified by adjusting the lengths so that it is clear that one side is shorter and the asymmetry can be read as an active choice.

4.6 ALIGNMENT

Sometimes elements don't line up neatly. As a general rule, it looks more orderly when elements are aligned. Sometimes the lack of alignment may seem necessary for reasons such as engineering or production. Bearing this alignment concept in mind, the designer may wish to see if some of the given constraints that lead to lack of alignment are as tightly fixed as when initially presented. Often non-design constraints are devised without reference to aesthetics, and when the outcome is identified, those responsible may be able to find a way to alter the constraints to help facilitate a more satisfactory design outcome. It is part of the designer's job to speak up for the user and bring their expertise with the visual aspects of a product to the attention of those whose concerns are directed elsewhere.[4]

In this example, we are considering a set of apertures or panels on an object. It is a hypothetical example with the problem plain to see. Real-world alignment problems will probably be a lot more subtle.

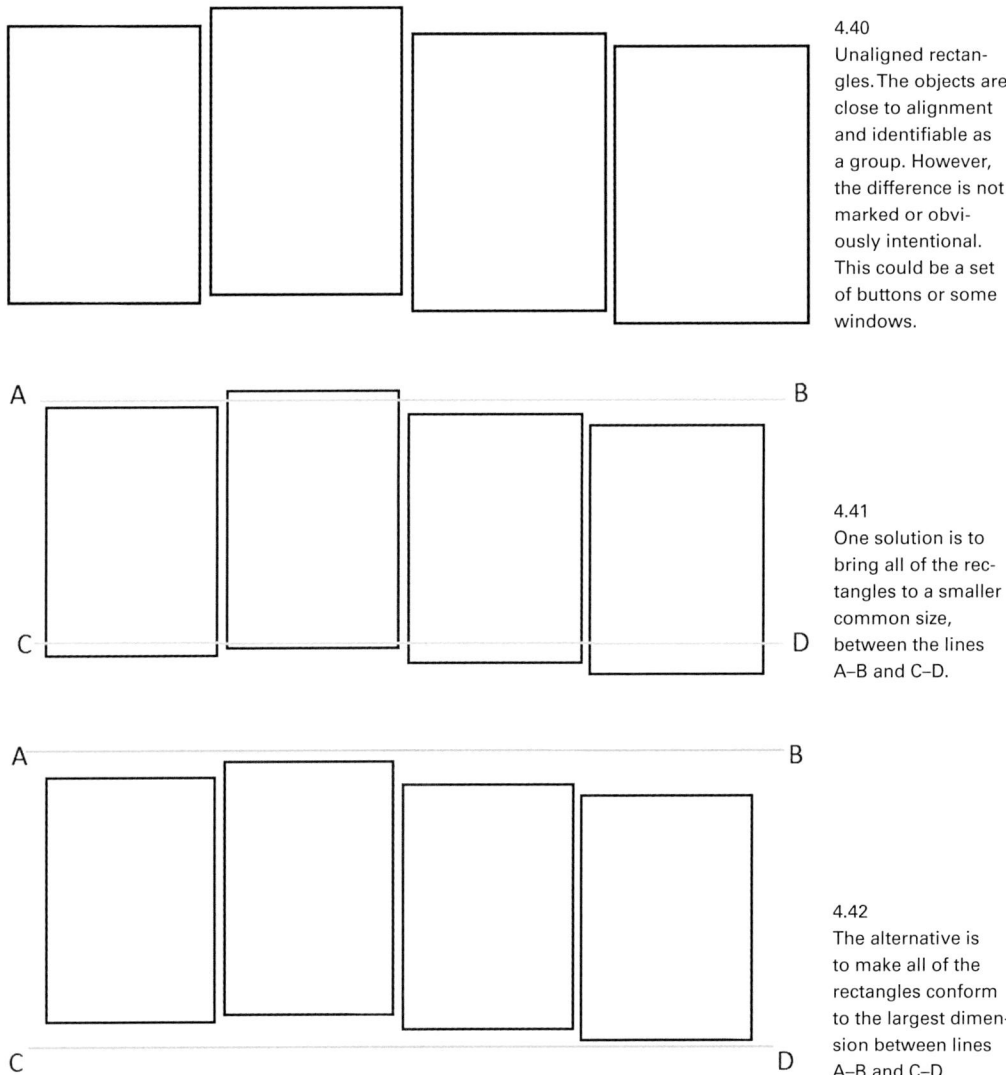

4.40
Unaligned rectangles. The objects are close to alignment and identifiable as a group. However, the difference is not marked or obviously intentional. This could be a set of buttons or some windows.

4.41
One solution is to bring all of the rectangles to a smaller common size, between the lines A–B and C–D.

4.42
The alternative is to make all of the rectangles conform to the largest dimension between lines A–B and C–D.

In the cases of both Figures 4.40 and 4.41 it assumed that there is no functional reason (from the user's point of view) to have differently sized units. When I am using this example in class, I typically present the rectangles as having some kind of a backstory: the location of the right-hand rectangle is there to suit the Canadian market; the other three are carryover locations from a previous design. I use this to explore the idea that an alert designer will (politely) inquire as to the reasoning for the constraints and then learn that the product is no longer on sale in Canada and that the carryover design is simply a default – that nobody actively wishes the elements to be positioned this way in particular. By uncovering the backstory, the designer determines there is no major reason for the exact dimensions and they can be revised to the advantage of the appearance of the product. Underlying this argumentation is that even if there are engineering requests for certain dimensions, most of them are of no interest to the consumer and, if necessary, can be changed up to the point at which the performance of the product is adversely affected.

4.7 ROBUSTNESS

Robustness has a number of dimensions. It's not just about being strong enough but *looking* appropriately strong too. This is related to the notion of semantics. Putting a high value on robustness is an old idea. Two millennia ago, the Roman writer Vitruvius (70 BC–c. 15 BC) summarised in his *Ten Books on Architecture* that good architecture required:

> *Firmitatis* (Durability)[5] – The object should be strong enough to endure.

This is true of design as well. We are concerned here not only with the fact of robustness or durability but the clear appearance of durability. Based on our experience as well as our intuition, we have an idea of what a robust object looks like. Of course, there may be artistic, creative reasons to play with notions of durability or to strive towards lightness for aesthetic reasons. Other things being equal, it is not desirable for an industrial designed object to look frail or wobbly. *Firmitatis* is inseparable from *utilitatis* and *venustasis*. *Firmitatis* also relates to proportions, as we will see below.

Generalising about robustness is not simple. The quality of robustness and the appearance of robustness dissolves in a cloud of subjectivity upon closer examination. All we can do here is to add another aspect of design to which the designer needs to be sensitive.

Consider the profile of something that looks like a bridge (Figure 4.43).

Geometrically there is nothing much wrong with this bridge shape, but the span could be said to be too thin in relation to the very thick pillars.

In the example in Figure 4.44, the connecting region has been thickened.

A case could be made that the solution in Figure 4.44 is too robust. If this is the case, the form could be fine-tuned to bring it into a better balance. That is a matter for further sketching and modelling, but at least the problem of the fragility has been dealt with. Often the difference between fragile and robust is a matter of a few percent.

4.43
The connecting part between the two vertical elements is mathematically adequate but looks frail in relation to the pillars.

4.44
A stronger-looking solution. There is no doubt this will carry people from one side to the other.

EXERCISE 4.5

The aim of this exercise is to become aware of the effect of small differences on how robustness is expressed.

Redraw an existing object such that some part of it looks fragile and frail. Then try to draw a version that is intermediate to the existing design and the revised design.

Redraw an existing object such that some part of it looks too massive and excessively robust. How easy is that? It might be rather hard to do so that the revised design still has any sense of being believable.

Apart from the shape of the object, how else is robustness conveyed?

4.8 CONTEXT

Deep immersion in the details of design can at times cause a designer to neglect to change perspective, to forget to zoom out and look at the bigger picture. Problems of context are avoided by remembering to take a break and look at it from a middle distance. Typically, a context problem boils down to having a detail that looks good up close but which does not fit properly into the overall whole.

This shape might look acceptable (Figure 4.45). It's a neat oblong with rounded corners. It could be the profile for a groove on a bigger surface, for example.

4.45
A nice, neat oblong with rounded corners. Modelling this with nice radii on the corners and making sure it's all aligned and tidy could take a while if it was set on a double curvature primary surface.

However, when set in the context of other features, the oblong is now quite obviously not consistent with the other oblongs (Figure 4.46).

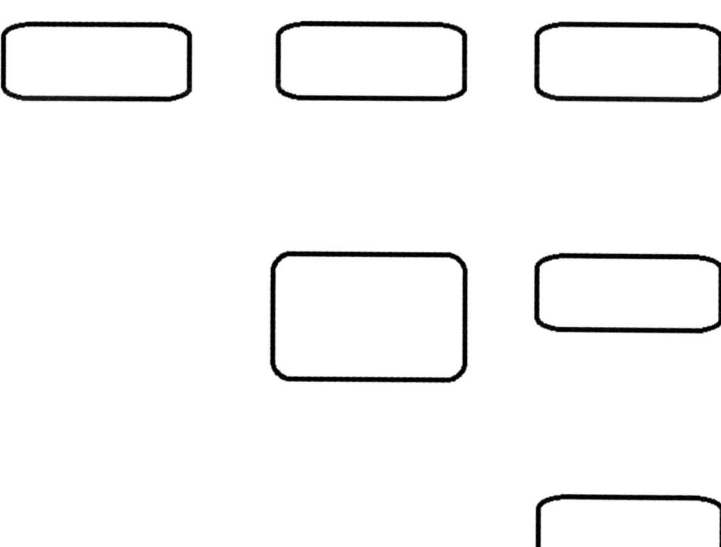

4.46
Seen in context, the larger form is not consistent.

For a more sophisticated example of the context problem, consider the door opening lever on this car door (Figure 4.47). Without a doubt, the geometry of the lever and its surrounding bezel is probably handled very well. There appears to be good lead-in on the corners and the metallic-effect lever was undoubtedly subject to intense consideration as surfacing/sculptural defects in such a prominent feature would be very unwelcome in terms of visual quality.

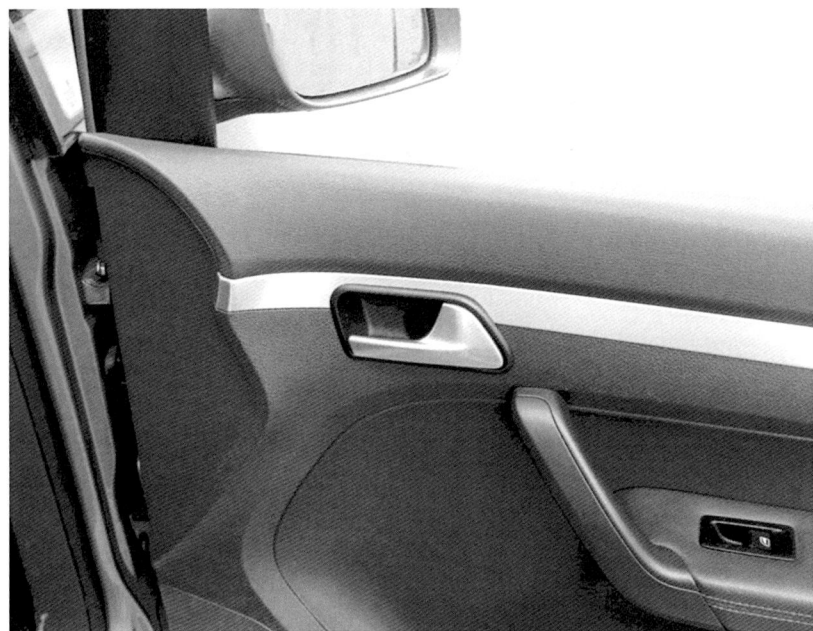

4.47
Door opening lever
on a car door. The
lever and its bezel
are neat bits of work.

It is when the door opening lever is analysed in the context of the overall landscape of the door that it seems not to fit in as nicely as one would like. On one level, the disinterested observer could say that the relation of the handle to the rest of the door is not glaringly bad. Many, in fact, would probably not notice anything amiss. Taking things for granted is not the hallmark of careful design, though. Without knowing the constraints, the door handle and its surroundings could be accused of the following inadequacies (see Figure 4.48, marked up). One, the gap between the lower right corner and the grab-handle is too small (A). It is not as you'd draw it. Two, the outline of the bezel surrounding the door opening lever is situated oddly in relation to the chamfer that runs below it (B). The bezel is also situated about halfway across a strip of metal-effect trim which does not look graphically robust (C). The overall effect of this is that the door opening lever's position has more than a touch of the accidental about it. There might be good reasons for this outcome, but the customer doesn't care about them; the customer sees a door opening lever that has a haphazard position in respect to the geometry around it. That is a context problem: fine door handle, bad placement.

To see this as a problem at all demands that the designer is concerned with avoiding giving hostages to fortune. Even if the user is not consciously aware of the shortcomings of a detail like the door shown above, they will be aware when a design seems pleasing or not. The coherence and internal consistency of a design go a long way towards providing a satisfying aesthetic experience. It might be that the user tries two products. One is like the car door shown

4.48
Door opening bezel and its surroundings, marked up. If you look at the image in a general, nonfocused way, the handle and bezel seem to float unrelated to the surrounding elements. The bezel seems to have been stuck on rather than being an integrated element of the door panel.

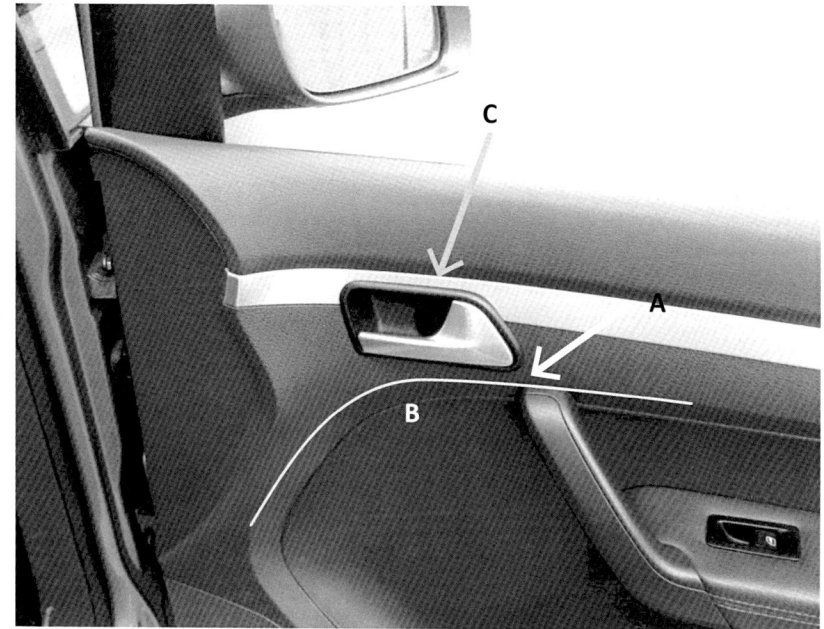

above, far from egregiously bad. The other does not have these shortcomings and registers with the customer as somehow more pleasant. If that one makes the sale, it's the design that withstands five or more years of continual scrutiny.

4.9 CONSISTENCY

Consistency is related to context but is more local. The underlying principle is to seek simplicity and the use of the same design rules as much as is possible with other factors. Another term for this is homogeneity, such that all of the smaller elements of a whole look like they belong together. Consistency does not mean uniformity; rather, it means visual interest within the framework of a few basic rules. A classical street is not typically uniform, but the buildings usually have much in common such as approximate heights, a small range of colours and some elements of style. The researcher R.A.G. Post (Post et al., 2013) made a similar point with reference to vehicle interiors, using the term 'unity-in-variety'. Assuming a smallish set of design rules for an object, the elements should be varied meaningfully. The viewer ought to be able to grasp the general theme and each element of the design should makes sense within it. The same applies to graphic design such that a layout is most easily understood when it's tending towards the restrained. In product design, consistency is needed among a product family, so the task might be ensuring that two or three products have a family look.

EXERCISE 4.6

At the end of this exercise you should become more aware of a product's form language and how it is consistently applied to a product (or not, as the case may be).

Find an existing object such as a chair, coffee maker, vacuum cleaner or even a work of abstract art and select an element for redesign. In order to do this, one must have a sense of the underlying principle being applied to the existing object. This is sometimes called a 'form language' but there is no grammar. It is more properly understood as a set of rules as to how to treat the main surfaces, how to treat edges and corners and the way the materials are deployed. There are other aspects to be considered such as proportion and surface treatment (rough, smooth, etc.).

The first objective might be to redesign part of the object in such a way that a disinterested observer will not be able to easily tell which is the original and which is the redesign when shown both.

You might notice that it is extremely difficult to alter a designed product (as opposed to an abstract sculptural work) since the constraints are so tight. That should tell you something about the degree to which industrial designed products conform to quite strongly applied rules.

In automotive design it is often felt necessary to partially redesign a car a few years after launch. The grille and front and rear bumpers might be altered and the metal panels carried over. Interestingly, few facelifts are especially successful. This is because the new elements do not have the same styling rules applied and the resultant whole is not as homogenous as the original.

For architects, the challenge in modifying a building is whether to be fully consistent or to add something markedly different. Either way, it helps to be sensitive to what is already there so that the intervention is a success. You could choose a building and draw two revised versions: one in keeping with the existing structure and one that is in sharp contrast. Discuss the results with your peers.

If you are in a graphic design mood, use the style of product to redesign the cover of a magazine. How convincing is the re-styled magazine cover? Or, conversely, find a news magazine and redesign a cereal packet using the magazine's fonts and colours.

4.10 DILEMMAS: WHEN NONE OF THE POSSIBLE
SOLUTIONS ARE ALL THAT GOOD

A dilemma occurs in design when there are two (or more) possible obvious solutions, each with apparent disadvantages. This is usually when several general principles are in conflict. In the case of the photocopier handle in Figure 4.21, the desire for a constant gap condition is in collision with the wish for symmetry.

There are larger radii on the outside side of the gap than on handle. Making all of the corner radii the same would mean two of them would be either wrong (too small or too big) or not work correctly.

At a workstation level, the dilemma is recognisable when you find yourself toggling between the two versions in a state of uncertainty. Using your own experience and judgement, you may find it difficult to assess which of the two options is the 'least bad'. Aesthetically speaking, you are in a state of ambiguity. At this point, the way out is to refer to intersubjective judgement (which solution offends users the least) and seek advice from colleagues or a higher level of management. But what either of these two approaches might show is that one of the parameters or constraints resulting in the proposed alternatives might be usefully reassessed if possible. Whilst an informed insider might understand the compromises necessary, the customer often doesn't. 'Why did they make it like that?' is a frequent cry of dismay from baffled users. The answer is often that the alternatives were worse. The customer doesn't know what they were, though, so all they are left with is an odd shape or ugly feature.

The aim should be to produce a form which makes sense without the customer or user needing to be aware of technical constraints over and above general knowledge. So, when you get to the state of a design dilemma, try to set down the drivers of the problem and evaluate which one needs to be revised in order to produce a clearly satisfactory result. With that knowledge in mind, a presentation of the unsatisfactory design alternatives can be discussed and a possible solution suggested: 'Here are alternatives A and B, both of which are inadequate in some way, but if we change factor X, we can produce a better solution. Is that possible, is that worth the effort?' This path is shown in Figure 4.49. It might come as a surprise that questions of aesthetics can be traced back to what seem like managerial advice. Designers need not be slaves to managerial considerations; being able to identify the managerial response to resolving an aesthetic problem is a good way to affect a better result.

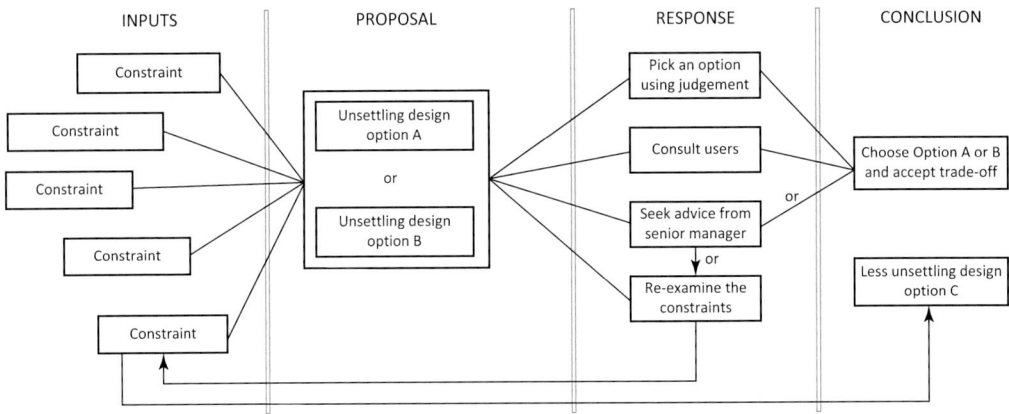

4.49
The path forward from two unsettling design options. Key here is that when you are confronted by two or more bad choices, don't be trapped into choosing one before examining the constraints more closely.

SUMMARY

After a lengthy tour through concrete examples, it is apparent that the principles outlined by Vitruvius two thousand years ago are still at play: robustness, utility and beauty. Easy to state and sometimes tricky to apply, these goals demand the balancing of requirements and constraints.

Another way of understanding some of these problems is to think about topology. Many designed objects are compounds of multiple parts, and where parts meet there are lines. So, many design problems arise from the need to apply or project lines onto positive (and negative) forms: a panel gap on a car, a joint line on a hairdryer or, more subtly, a tool-split line on a plastic part.

At a general level, this chapter makes clear some of the common problems encountered in terms of a set of cases. The overarching point is that the designer needs to be self-aware and step back from the problem to assess it. If sketching and CAD modelling fail to provide an easy-enough answer (when you have been modelling for several hours without success), it is time to evaluate the input, the constraints and the goals to see which of those needs reconsidering. Perhaps a required constraint is not as sacred as was initially thought. With the best will in the world, no engineer or product planner or designer can know in advance which requirement A, B, C, D or E, etc., should be compromised at the end of the design process to achieve a satisfactory result. Vexing as it might be to reconsider a requirement – and, indeed, it might require more work to adjust other perhaps invisible elements – it is better than presenting the customer with a baffling shape.

Consciously or unconsciously, ambiguous shapes and strange forms act as reasons not to buy or not to retain the item. If these reasons can be minimised or eliminated, the designer is on the path to creating an object the user is pleased with and will keep for a long time.

NOTES

1 The literature on joints is scant. While writing this I was offered a variety of references by colleagues, and almost none of them offered much of any use. Even this Frampton quote is on the margins of uselessness. What I have written myself on the topic of joints in architecture is what I expected to read in Frampton and did not find. So well done, me.

2 I look at this joint on my commute to work and have concluded that to avoid this butt joint would involve a dramatic reworking of the train's interior panel. This was either too costly to undertake at all or the problem, such as it is, turned up too late in the design process to rectify.

3 This is an example of a design dilemma. The wish for visual consistency suggests having similar radii on both side of the gap. The wish for a smaller apparent gap suggests eliminating the radius or minimizing it.

4 'Industrial designers and engineering designers do not consider the same aspects in the industrial Product development work. Industrial designers focus on social utility values, such as aesthetic, appreciation, emotions and understanding, while engineering designers usually concentrate on material utility values' (Muller, 2001, as cited in Persson and Wickman, 2004, p. 1).

5 And also *Utilitas* (Utility) – It should be work well and be useful; *Venustatis* (Beauty) – It should look lovely, raise the spirits.

FURTHER READING/KEY TEXTS

Author's note: The academic literature on craftsmanship is scant despite the importance of craftsmanship to perceptions of quality and aesthetics. I have included David Pye's book because at a general level it is very helpful. It is also a delight to read. It is, however, short on specifics.

Frayling, C. (2011) *On Craftsmanship*. London: Oberon.

Holl, S., Pallasmaa, J., & Perez-Gomez, A. (1994) *Questions of Perception: Phenomenology of Architecture. A+U*, Special edition 7.

Persson, S., & Wickman, C. (2004) Effects of industrial design and engineering design interplay: An empirical study on tolerance management in the automotive industry. Paper presented at Paper presented at the International Design Conference – Design 2004, Dubrovnik, Croatia, May 18–21.

Post, R.A.G., Blijlevens, J., & Hekkert, P. (2013) The influence of unity-in-variety on aesthetic appreciation of car interiors. Paper presented at: IASDR 2013, Tokyo. http://design-cu.jp/iasdr2013/papers/1630-1b.pdf.

Pye, D. (1968) *The Nature and Art of Workmanship*. Cambridge, UK: Cambridge University Press.

5 The meaning of the object and its elements

Product semantics

5.0 INTRODUCTION

Jeffrey has a job interview with a prestigious investment consultancy. For the big day, which is a warm one, he decides to wear a sports t-shirt with a large numeral on the front and a pair of linen cargo shorts. Sandals complete the look. All in all, the ensemble works in terms of function and colour coordination. He does not get the job.

Jane is an MA student who has designed a hospital complex for her final project. She has researched the technical demands of a hospital and conducted interviews with staff about the wards, the staffrooms and patient waiting areas. Referring to several major technical studies, she is confident the spatial allocations are good estimates. Finally, for the exterior Jane has chosen matte black stone, dark grey concrete, mirrored glazing and black anodised metal panels for the façade. The main entrance is a nine-metre-high doorway in black marble. It is approached down a long, deep walkway defined by four-metre-high rusted sheet steel with jagged upper edges. Jane's grade is a disappointment.

Jeremy is a graphic designer. He has devised a poster which is intended to advertise outdoor holidays in the Italian alps. He has checked that the font sizes are legible to all user groups and that the photo images are up to date along with the facts in the text. For the layout he has chosen red-grey, black and grey-green. The font is a very ornate one such as Gloucester MT Extra Condensed and the text is arranged in 1960s-inspired wavy lines. The client is not impressed.

In all three cases, the geometry and facts of the designed objects are not problematic in themselves. The proportions, curvature and craftsmanship are beyond reproach. What is problematic is the *symbolism* of the objects and their elements. It is a matter of a failure to pay attention to the semantics of the design tasks at hand. So, how to avoid sad (and extreme) cases like this? In this chapter we shall take a look at the meaning of form. This will allow you to work forward (from meaning to shape) and to look back (from shape to its possible meaning).

The short, graphic answer to the question 'why product semantics?' is shown in Figure 5.1.

DOI: 10.4324/9781003183303-5

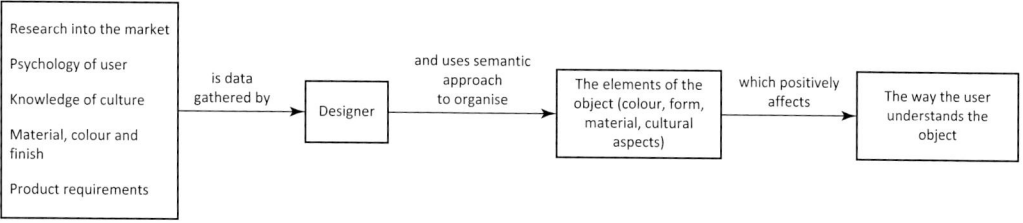

5.1
Why product semantics?

In this chapter form will be considered on the basis of a symbolism such that it can be interpreted as a whole and in parts. Designed items can be understood as having features which have potential symbolism. This is known as the *semantic approach* to design, which is 'how people attribute meanings to artefacts and interact with them accordingly' (Krippendorff and Butter, 2008, p. 2). The study of product semantics is about understanding the interpretation of designed objects.

We can use the metaphor of language to think about the appearance of objects. A real language is a consistent set of rules and parts of speech controlled by them. A product *language* mediates what a product is intended to express (Gros, 1983) in a way analogous to language being a medium of thought. The product language is conveyed through characteristics such as 'form, dimensions, colour, graphics, texture, transparency, groupings of product parts etc.' (Boess and Kanis, 2008, p. 307). The aim of using product semantics is to facilitate communication (of the product's message) between the product and the consumer (Vihma, 1995). Product semantics are closely related to the concept of style. When we talk about Givenchy or Doc Martens having a style we are touching on the semantics of those brands.

Before diving into the topic, we need an overview. One way to relate the elements of this chapter is shown in Figure 5.2.

For this chapter I take an agnostic and pragmatic approach to the concepts related to product semantics. This is in order to present ideas which may work as useful rules of thumb without accepting them wholesale. As with all such ideas, a degree of scepticism is healthy. Scruton (1979) and Weber (1995) presented quite substantial critiques of the thinking behind product semantics. Those critiques rest on scepticism about the extent to which designed objects can be said to have grammar and vocabulary. It is not the aim of Scruton or Weber to deny that people (and especially designers) may view objects from a semantic perspective. It is rather to point out the limits to this way of seeing. I will address those at the end for the sake of balance.

Because of the complexity of cognitive processes, there is some overlap with between semantics and aesthetics.[1] Whether something is perceived as attractive has at least something to do with whether it's meaning is in accord

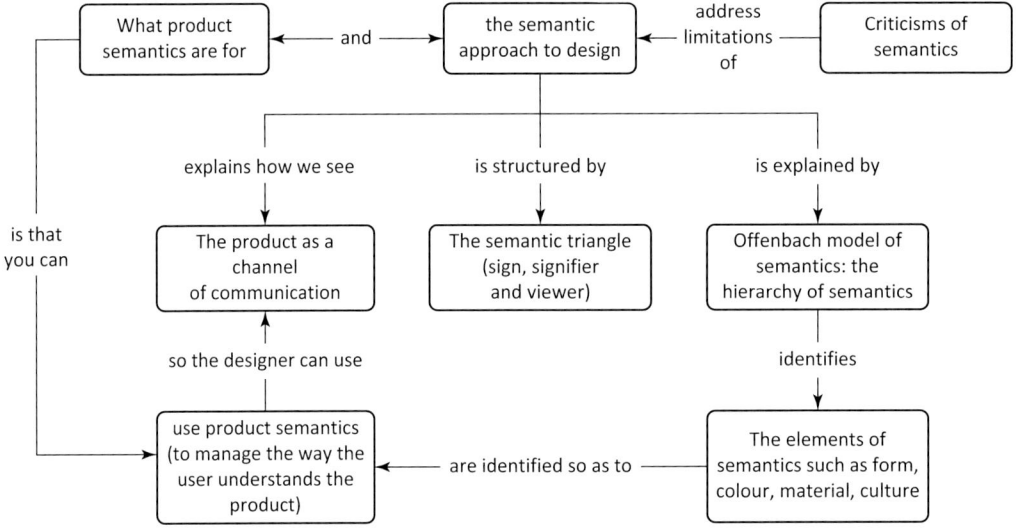

5.2

How the elements of this chapter are related. The central part seems to be the idea of the product as a channel for communication. Successful products have a clear message for the user.

with its function or vice versa. The cognitive and sensory explanations of visual perception (Chapter 1) and how an object is interpreted symbolically are related. That is why the presentation of reading a form 'as if' appears in Chapter 1, Section 1.6, when it could also be presented in this chapter as a type of symbolism. That way of looking is based, I would contend, on the viewer's learned experience and haptic sense rather than being *primarily* culturally derived. However, since cultural values are based in part on the facts of physics and biology, the cognitive and semantic are inevitably hard to completely separate.

5.1 WHAT IS THE SEMANTIC APPROACH TO UNDERSTANDING DESIGNED FORMS?

This chapter examines the way forms and elements of form can be seen as signs or to have a meaning beyond their pure geometrical content. This is relevant in the phase when a designer wants to generate shapes and in the phase when the designer wishes to assess the forms she or he has created. In very simple terms, objects can be interpreted as having some kind of meaning, in whole or in part. In this view, the object as whole may be read as a sign of something else and its parts may also be read as a sign of something else. This is shown in Figure 5.3, which shows two garments for use in inclement weather. They differ not in their functionality but in their meaning. The elements of the coat on the right have meaning and the garment as a whole has a meaning: a modern coat for a modern person. The semantics of the coat on the right are chosen by the designer to convey modernity. The coat on the left is also a functional garment. It conveys notions of continuity, tradition and robustness. Although some users might be indifferent to the coats' meaning (or image), most would not be and would have active preferences for one or other item due to the semantic content. We need to take a closer look at the term 'semantic'.

5.3
Two raincoats. They differ in their use of product semantics. Which one is for traditionalists? (Left: courtesy of Barbour; right courtesy of WantDo Inc.)

'Semantics is the study of the meaning and reference of linguistic expressions, while semiotics is the general study of signs of all kinds and in all their aspects' (Føllesdal, 1997, p. 449). Semantics is nested under the general category semiotics. 'Product semantics is the study of the symbolic qualities of man-made forms in the context of their use and the application of this knowledge to industrial design' (Krippendorff and Butter, 1984). The authors of that article qualified that remark as follows: 'We take semantics as the study of meanings in the broadest sense, not to be confused with how the word "semantics" has been appropriated in the rigid structures of semiotics' (Krippendorff and Butter, 1984, p. 4). Some authors talk in general terms about a *semiotic* approach to design (e.g., Vihma, 1995; Monö, 1997) and some talk about *semantics*. Apart from a few instances of concern to professional philosophers, the distinction is not important for our purposes and I will use the term 'semantic' as much as possible throughout this chapter. You can fairly harmlessly assume that product semantics covers the way in which a style is interpreted and the elements associated with style. More so than pure geometry or shape, semantics is quite wide open to interpretation. Whilst there might be quite broad agreement on the nature of a curve, there is likely much less about semantics such that one person's convincing design might for another person be a case of unsettlingly bad taste.

EXERCISE 5.1

The aim of this task is to explore in a classroom setting notions of style and bad taste. You are unlikely to reach consensus on this.

Each individual in the class should find an example of something they'd call stylish and something they might identify as bad taste. Try to pin down why these are cases of stylishness and bad taste. What is behind these diagnoses? What elements are most strongly associated with the style? And what elements are contributing to the bad taste? Put another way, what would you change to correct the perception of bad taste? The bad taste example will be easier to work with than the stylish example.

Designers are known for their views on aesthetics and style. You might want to go beyond the view of your classmates and teachers. So, a good additional exercise is to select the examples of bad taste about which the class most agrees and ask some laypeople what they think. Do the same for the case of stylish design? Take a note of the participants' comments. Are they in-line with the class's views?

5.2 SIGN, SIGNIFIER AND VIEWER: THE SEMANTIC TRIANGLE

Behind this notion of semantics is the relationship of the object, the thing it refers to and the viewer seeing the object. It is a three-way relation and includes a considerable degree of subjectivity. Figure 5.4 shows what is called the semantic triangle. The concept first appeared *in The Meaning of Meaning* in 1923 (Ogden and Richards, 1923). A very simple example would be a no smoking sign. This case consists of a relationship between the viewer and the object in a three-way structure. There is

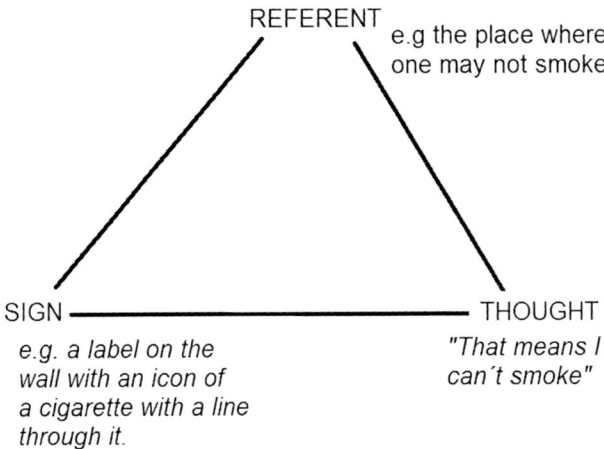

REFERENT
e.g the place where one may not smoke

SIGN
e.g. a label on the wall with an icon of a cigarette with a line through it.

THOUGHT
"That means I can't smoke"

5.4
The semantic triangle (adapted from Krippendorff and Butter, 1984).

1. the thing ('referent'),
2. the sign indicating the thing, and
3. the thought about the thing.

I will return to this semantic triangle later on in the chapter when this relation is used to understand three-dimensional form or details. At this point I will show how an object as a whole would fit into this framework using the Fukasawa kettle as a case object (Figure 5.5). Remember, although the kettle is meant for heating water, it is also acting as a sign. The kettle in *your* kitchen is a sign, too.

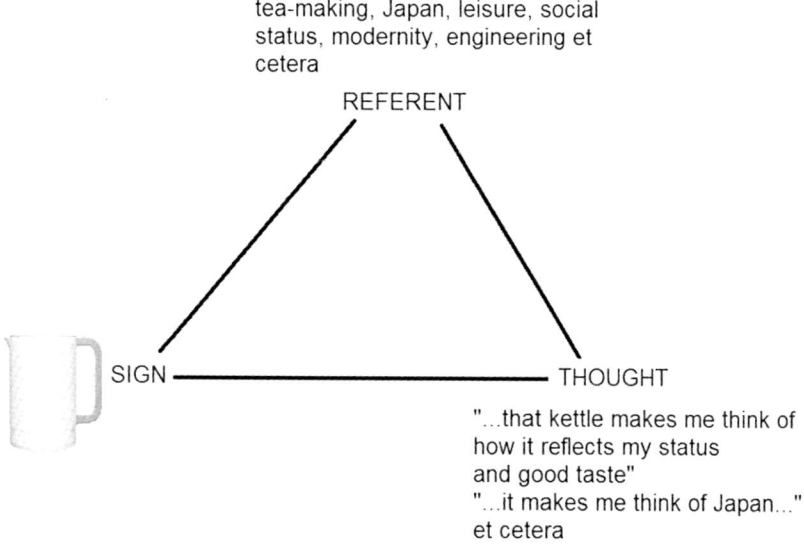

tea-making, Japan, leisure, social status, modernity, engineering et cetera
REFERENT

SIGN ——————————————————— THOUGHT

"...that kettle makes me think of how it reflects my status and good taste"
"...it makes me think of Japan..."
et cetera

5.5
The product as a sign.

The difference between the case in Figure 5.4 and the case in Figure 5.5 is as follows. In the applied case (Figure 5.5) the sign has a use beyond being a sign (it boils water, as I said), whereas in the simple case this part of the relation, the graphic, is an image of a crossed-out cigarette (as in Figure 5.6). The graphic is only a sign of something else and no more. It would possibly be a sticker on an airplane seat back or a plastic sheet attached to a wall. It doesn't do any non-smoking itself. What it does is act as a label.

Krippendorff and Butter (1984), who brought product semantics to broad attention in the 1980s, claimed that the semantic triangle is not directly applicable to the world of designed objects. This is because an object's form 'says something about itself, and second something about the larger context of use and both about the user who interacts with it and makes the conceptual connection' (Krippendorff and Butter, 1984, p. 4). They went on to add that 'an object's form does not say what it is' (Krippendorff and Butter, 1984). That is true in a literal sense since the kettle in our case cannot speak to declare 'I am a kettle' but, by the same token, the 'no smoking' sign does not say what it

5.6
'No smoking' icon.

is either. Both the kettle and the sticker need to be *read*. The object is what it says to the user and it expresses this by means of associations. More precisely, the designer has managed the form to guide the user to desired associations.

Chapter 1 explains one way of 'reading' the kettle or any other object because the form of the object reveals something of what it can do (its affordance). The kettle has a handle, which means it can be gripped, and it has a locally curved surface indicating that a liquid could be poured from it. Whereas Krippendorff and Butter (1984) reworked the semantic triangle based on their notion that the object does not refer to things, I contend that the semantic triangle can be seen as a useful conceptual shorthand. In rough and general terms, design is a form of communication even if the parallels to human verbal and written communication are not complete or direct.

EXERCISE 5.2

The aim of this exercise is to imbue an object with meaning without relying on clues that suggest a practical function. Practical functions are left out to focus on other aspects of form as semantic signals. Sculpture is the shaping of form to make associations either directly through representation – for example, a human figure – or indirectly with abstraction.

If, at least in metaphorical terms, objects can be read, then they can communicate something. Take as a basis a plain cube (like the sketch below) and attempt to reshape it so that it expresses a message (broadly understood). The redesign should still retain enough characteristics of a cube to be seen as a modified cube and rather than a thing with some vestigial cubic character.

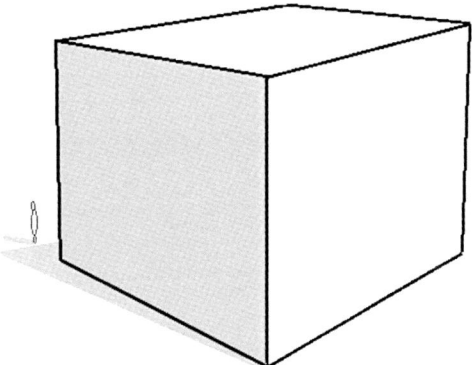

When you have reworked the cube in as many ways as possible, try to classify the ways the cube has been reworked. For example, if you have redesigned the cube so it is covered in hand-stitched leather, then the reference might be (albeit vague) to expensive shoes, handbags or saddlery. Another cube could be fluted and refer to columns used in classical architecture. A third cube could be smoothed off and resemble some of the equipment found in a dental surgery.

5.3 WHAT PRODUCT SEMANTICS ARE FOR

Most designers work with product semantics at an intuitive level, selecting shapes they think look right and which convey a general sense of the object to the consumer. Consumers themselves have a grasp of this and may accept or reject objects on the basis of product semantics when, say, choosing a matte black television instead of a glossy white one or a pair of Dr Martens instead of desert boots. Designers use product semantics to guide the customer to the product and its values.

Product semantics can be used to bring the user from one technology to another. The use of familiar icons (the rubbish bin icon) on the graphical user interface of Apple computers is an example. We know how a bin works and the icon suggests that unwanted files can be disposed of. The conventional form of the unconventionally powered Tesla S electric car would be another example. The Tesla is a case of deliberately *not* signalling high-tech but instead reassuring normality. The Tesla looks like a car with an internal combustion engine and has the form language of a vehicle with a motor under the long bonnet. However, the technology of electric cars does not require distinctly separate volumes for each function. Conversely,

> product semantics can emphasize technical progress by giving products an eye catching, unfamiliar or futuristic appeal. ... Alterations of basic societal

conditions, socio-cultural upheaval, and the arrival of new viewpoints and values are just as important for the creation of innovative designs or new product languages. Artefacts that are not innovative with regard to technology might be epochal with regard to the expression of the spirit of the time.

(Steffen, 2010, p. 85)

With that in mind, product semantics can be used to signal other emotional and cultural values and can help reposition the product in a new way.

The next item to consider is where in design it is helpful and necessary to consider product semantics.

In one of the most influential texts on the subject, 'Product Semantics: Exploring the Symbolic Qualities of Form' (Krippendorff and Butter, 1984), there are four categories where 'meanings relevant to product design' can be considered. These are:

1. information displays
2. graphic elements
3. product's form and texture
4. indications of the product's internal state.

The authors emphasise 3 and 4 as being the most important. The product's form and texture 'cannot be understood without the object being that object for the user. The symbolic meanings of forms, shapes and textures are the most characteristic concern of product semantics' (Krippendorff and Butter, 1984, p. 6). The indications of the product's internal state include features such as the position of switches showing whether the device is on or off, a lever in one or more different positions or the positioning of components that might be in use or disconnected. The third category touches on affordances. Krippendorff and Butter's examples of 4 refer to things like dial needle positions and flight instrument indications. They overlap partly with 1 and also relate somewhat to perceptions of affordance.

Regarding information displays, Krippendorff and Butter (1984) explain that the designer's concern is with the interface between the user and the displays and not the content. Except when used decoratively, the written graphic elements have their own semantic domain (an area of linguistics). Thus, of the four categories listed by Krippendorff and Butter, only one unambiguously stands out as being related to product semantics, item 3, form and texture.

Semantic modes of thinking often operate at a subconscious level. This means that if one asks a designer to sketch a proposal for a piece of furniture or a toaster for a certain user the designer will tend to choose some forms over others. They may not be fully aware of the reasons for the design choices made. They are not arbitrary, though. One might expect a running shoe designed for fashion-conscious middle-aged men to look somewhat different to one designed for committed athletes aged 18–30. The forms of the shoe proposal will have a distinct character. The particular curves, colours and lines add up to a shoe that signals its purpose to the desired customer.

5.7
Portable radios/hi-fis. Denver DAB 38 (left) and Lenco SCD 420 (right). (Images courtesy of Denver and Lenco)

Consider the radios shown in Figure 5.7. Each of these portable radios/hi-fis has a form organised to convey meaning to the user.

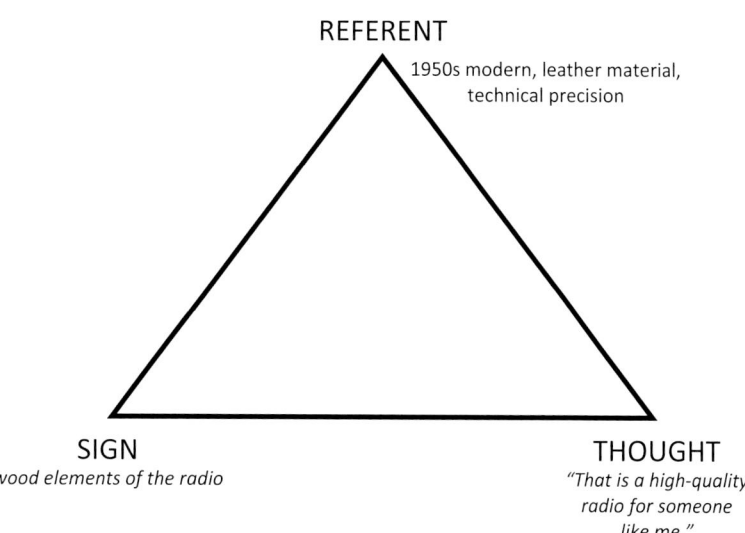

REFERENT
1950s modern, leather material, technical precision

SIGN
wood elements of the radio

THOUGHT
"That is a high-quality radio for someone like me."

5.8
The semantic triangle and material applied to the radio.

In comparison with the layout of the Denver radio, the pairs of the three small buttons on the Lenco SCD 420 could perhaps be called arbitrary despite being symmetrically placed around the vertical axis. The use of chamfers on the upper corner of the CD lid may arguably be inconsistent with the overall theme of large radii and roundness. The Denver is overall a calm and considered item, blending modern precision with retro touches. The Denver radio is analysed using the semantic triangle in Figure 5.8.

Product semantics does not relate directly to affordances (what the object looks like it can do) but rather the symbolic content: 'The symbolic meaning of form, shape and texture are the most characteristic concerns of product semantics' (Krippendorff and Butter 1984, p. 6). The author Möno (1997) took a diametrically opposite view, holding that affordances matter to semantic interpretations. The reason this is possible is because affordances unavoidably have some semantic content: the handle can be said to symbolise gripping, for example. However, the handle-ness of the handle is its most important quality. Its purpose is not to symbolise but to allow the object to be grasped. Semantic content could, in theory, be removed to a very large extent without affecting the immediate functionality of the object.[2] A good way to distinguish affordances and semantics is to ask whether the element is primarily functional or whether it is primarily suggestive of something else. In some cases there will be no clear distinction – it will be a bit of both.

Product semantics are also not about applied graphics. However, the graphics *as abstract fields* of colour related to the overall form have a semantic interpretation. The use of lots of labels, markings and decals on 1980s Japanese radios was not only about informing the customer of the functions but to suggest that the product was lively, exciting and in tune with the styles of the times. The form of the words mattered more than their literal meaning. The meaning of a large brand logo on a t-shirt front is different from the same logo embroidered with small lettering on the upper left or right side. Both say the shirt was made by X, but the large logo is also saying 'Look, I have a t-shirt from X'.

If the product is the means to communicate something to the user, the design of the product ought to be arranged to reduce confusion (noise) and make the message clear. The radios in Figure 5.7 have much the same function but are given forms that attract different customers. The customers will have consciously and unconsciously 'decoded' the products based on their appearance. The products are part of a communication pathway between the designer and the customer. That pathway can be visualised as shown in Figure 5.9.

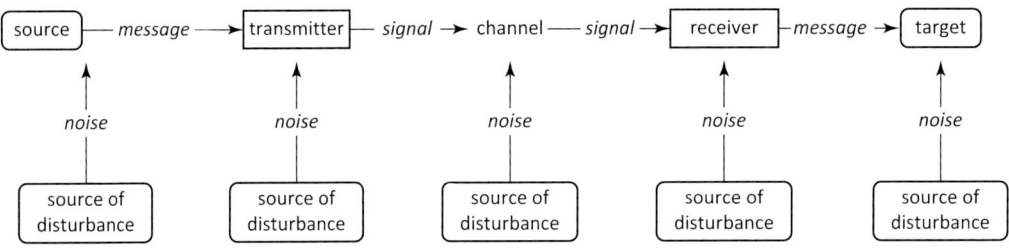

5.9
This diagram also shows where confusing signals can come in. (Adapted from Möno, 1997)

In Figure 5.9, the 'channel' is the product, such as a radio, chair, shoe or toothbrush. And this simple example shows how it can all go wrong for, say, another kettle (Figure 5.10).

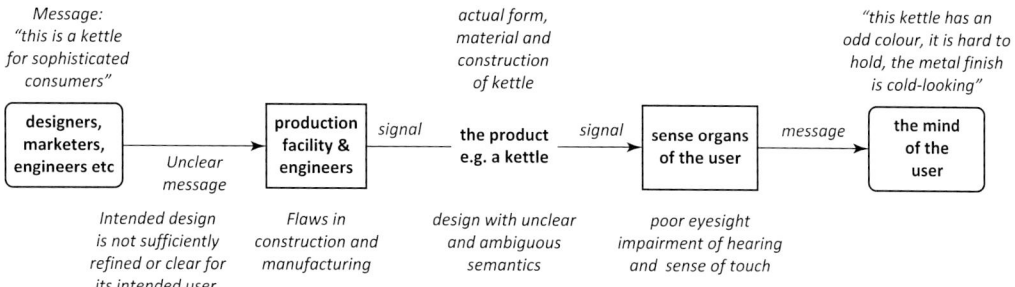

Message:
"this is a kettle
for sophisticated
consumers"

actual form,
material and
construction
of kettle

"this kettle has an
odd colour, it is hard to
hold, the metal finish
is cold-looking"

designers, marketers, engineers etc → *Unclear message* → production facility & engineers → *signal* → the product e.g. a kettle → *signal* → sense organs of the user → *message* → the mind of the user

Intended design
is not sufficiently
refined or clear for
its intended user.

Flaws in
construction and
manufacturing

design with unclear
and ambiguous
semantics

poor eyesight
impairment of hearing
and sense of touch

5.10
Shows the ways in which a product message gets lost along the way from the first idea to production. (Adapted from Monö, 1997)

The part we are interested in is how the message about the kettle (starting on the left of the diagram in Figure 5.10) turns into the product. If the semantics are unclear or plain wrong, the user will not receive the intended message.

The general gist of this approach is that 'an implement's design can signify something other than its original purely useful function' (Monö, 1997, p. 11). Vihma (1995) wrote that 'in addition to their practical functioning, they [products] also have an ideological role. In their ideological function products are signs that refer to something other than the material product and its practical functions' (p. 12). As an example of this and referring to the raincoats in Figure 5.3, the ideological role of the raincoats is to relate the user's property to themselves ('this is the raincoat for someone like me'). The coats also relate the users to their social group and to society at large ('this raincoat represents me and my values to others'). Consider the kettles by Naoto Fukasawa (left, 2008) and the anonymous one (right, 2018) in Figure 5.11.

5.11
Kettles from Japan (left) and the United States (right). (Left image courtesy of Naoto Fukasawa Design Ltd. Photo by Hidetoyo Sasaki)

These kettles are perhaps more subtle examples of the ideological role of a product and how it is not always uppermost in the consumer's mind. The way in which these kettles is viewed depends, of course, who is viewing it. The designer should be prepared to analyse these things in relation to as many cultural references as they can think of. This is despite the fact some customers may see nothing but one among many nice white kettles.

EXERCISE 5.3

In order to keep the consideration of semantics linked to real material things, this task is focused on tangible geometry of the kind considered in Chapter 3 plus other elements. This exercise should make you aware of the ways things can be interpreted.

Product semantics are ultimately tied to individual curves, surfaces and lines plus colour and texture. Using Figure 5.11, analyse the number and type of geometrical elements in each kettle. What do you think the elements might be trying to achieve? What about the way the forms are separated or blended? That has semantic content as well. The kettle in Figure 5.11 is possibly referring to the designs of other kettles to appear as a contemporary design.

5.4 THE CHANNEL: BREAKING DOWN THE ELEMENTS OF SEMANTIC PERCEPTION

Having considered in general terms the concept of the semantic triangle and how the designed object carries a message or signal to the user, it is time to disentangle the various elements of the object as seen from the user's eye view. To achieve this we need first to recall Figure 5.9, which shows the signal pathway. In the middle of the diagram is the 'channel', which is the product: washing machine, chair, car or hairdryer. The following diagram (Figure 5.12) shows Figure 5.9 reduced down to a small icon, and in the middle part the channel is expanded out into more elements.

This diagram serves as a good place to start in considering the parameters influencing the way the object is perceived. The main part of Figure 5.12 shows the various elements of which the *gestalt* or total perceived object are composed. This is Kicherer's (1987) conception of the elements of what the viewer sees. In Kicherer's conception the outer surface and coverage of the object are insufficient to describe the totality of the experienced phenomenon. Kicherer includes notions of the manufacture and of construction (e.g., ideas of plastic moulding), notions of what the item is for (you will judge a

5.12
In this diagram the elements of the 'channel' from Figure 5.9 are expanded out. The main part of the diagram (adapted from Kicherer, 1987) shows all of the elements that add up to a perception of form. The 'gestalt' is the whole object (from the German word for 'a whole entity').

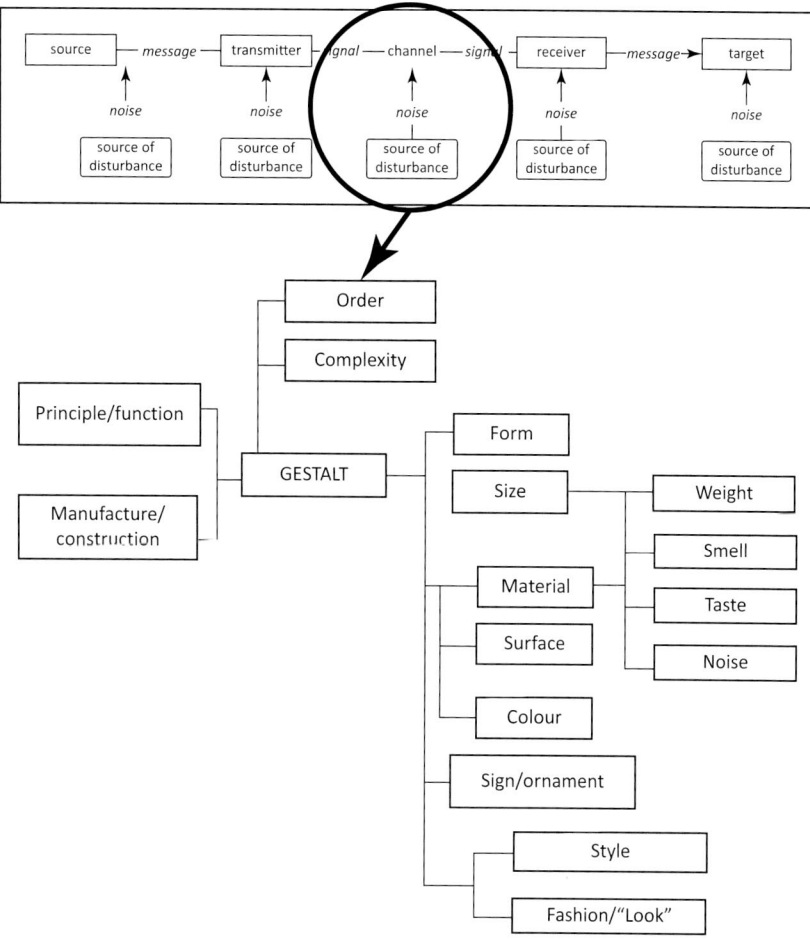

car depending on how it is presented, for example) and the degree to which order and complexity are in balance. All of that adds up to a perception of the entire object. To underline that distinction: for Kicherer form is one of *three* ingredients in the total perception of the object, the 'gestalt', rather than the only one.

- Form – how it looks.
- Order/complexity – is it simple or complex or a blend?
- Construction and function – how it is assembled to do its job.

What Kicherer (1987) does with this analysis is to usefully highlight the fact that the way form is interpreted is connected to an understanding of what the object is for and how it is made (resonating with Chapter 2, Section 2.4).

Although Kicherer's (1987) conception raises some questions about the boundary of the object, it gives an overview of the elements one must be aware of when analysing the totality of the user's semantic perception. Of interest to the reader will be Kicherer's placement of 'style' and 'fashion/look' in separate boxes *parallel* to the other inputs. Take another look at the lower right corner of Figure 5.12. One could easily consider 'style' and 'fashion/look' as being the same. And one could also consider the 'style' as being a particular combination of geometry, material and colour rather than a characteristic separate to them. To resolve this uncertainty one could remove 'style' from the diagram or place it elsewhere. Figure 5.13 shows an alternative way to think of the hierarchy.

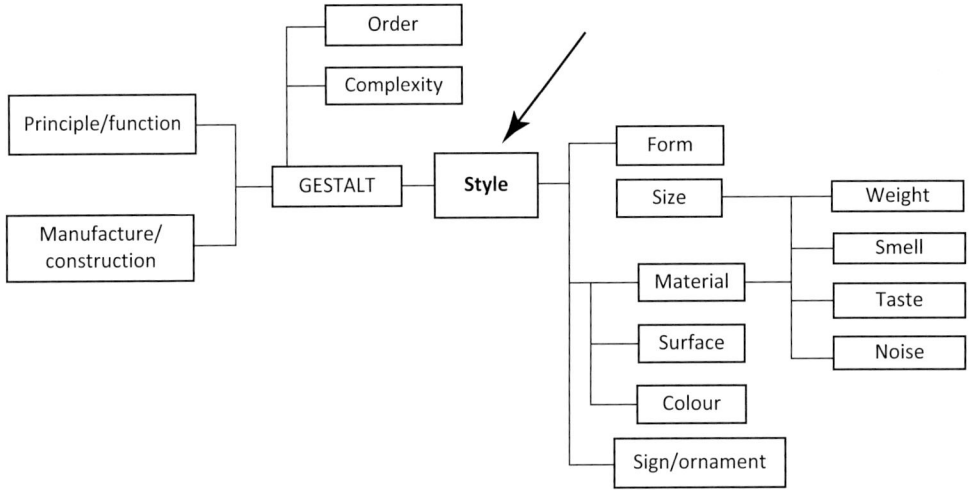

5.13
The relation of style to semantics in a revised version of Kicherer's conception of product semantics. Style is now an umbrella term for the elements connected to its right.

EXERCISE 5.4

The aim of this activity is to think about style as a whole. It can be something one aims at achieving in the beginning of sketching or the seemingly inevitable result of other inputs.

Discuss what style might be with your classmates or write a short paragraph about it (150 words). Write up a list of the factors that make up what is generally considered a style. How does the style of a product, building or clothing relate to meaning? Consider your own clothes. They probably have a style. What is the intended message and what physical elements contribute to it?

EXERCISE 5.5

The aim of this task is to use visual materials to dissect style. This is what you would do when trying to understand a products relation to other ones. It is also helpful in creating visual references for inspiration and analysis.

Assume 'style' and 'fashion/look' are much the same thing. If style is composed of materials, colour and particular geometries, select images that represent the style of either a time period; for example, the 1990s, the 1970s or even 1920s. Or select images for a contemporary 'style' of product; for example, a selection of modern products with a common style. Create a mood board showing exemplars of the colours, the material and the geometries. Clearly label the focus of each grouping. There will be some overlap between colour and materials. Since we are interested in semantics, or the meaning of these images, try to discuss and then decide upon possible meanings of the colours, material and geometries. Can you identify and argue for a general theme to the mood board? If you have mood boards for different time periods, can you compare them to identify their intended character? What were the 1970s or 1980s about, for example?

EXERCISE 5.6

The aim of this task is to use the conceptual structure in Figure 5.13 to analyse a product.

Find an item and analyse it using the categories in Figure 5.13. You could choose two items from the product group – for example, two sets of running shoes, two pairs of underwear or two armchairs – and see how they differ. The end result is that you should be able to express what the products are intended to mean to customer. You should also be able to use Kicherer's concepts as an analytical tool.

How does this structure apply to graphic design? Does it translate? How does it apply to fine art; for example, a sculpture? Does it work for premodern designs such as antiques? If not, why not?

Colour, material and finish (CMF) are important parts of design and formgiving. Can the semantic model be applied to the rather abstract subject of CMF?

5.5 PRODUCT SEMANTICS AS A DETAILED HIERARCHY OF ASSOCIATIONS.

For an even more detailed analysis of semantics I will present the Offenbach theory of product semantics. This section is indebted to Steffen's (2010) lucid summary of the subject. In the 1970s and 1980s Jochen Gros of the Academy of Art and Design, Offenbach, Germany, developed a fine-grained and systematic theory of product semantics which differs from Krippendorff and Butter (1984) by its focus on the object (item 3 in Krippendorff and Butter's schema discussed in Section 5.3).

The Offenbach model 'subdivided the specific object of product language into formal aesthetic functions, i.e. those aspects that can be observed irrespective of the meaning of their content – and the semantic functions' (Steffen, 2010, p. 87), which are open to interpretation. In this model Gros 'makes a distinction between the practical functions of a product (and various others such as ergonomic, economical, ecological functions) on the one hand, and the formal and communicative aspects, the so-called product language functions on the other' (Steffen, 2010).

The following three diagrams break down the Offenbach system into its component parts. The reader should keep in mind that these elements are ultimately related to the geometry, material, colour and texture of objects and are not intended as academic abstractions.

Figure 5.14 shows the overall structure of the Offenbach model. In this diagram, the term 'function' refers to an aspect of the object with an associated intention: something the object does or something the object is for. The function is conceived as being intermediary between the user and the product. Functions are subdivided into 'product functions', which would be those elements related directly to a task; for example, toasting bread, operating a servo motor or adjusting system states. The 'product language/sensual functions' refers to the nature of the object's geometry (more or less what it looks

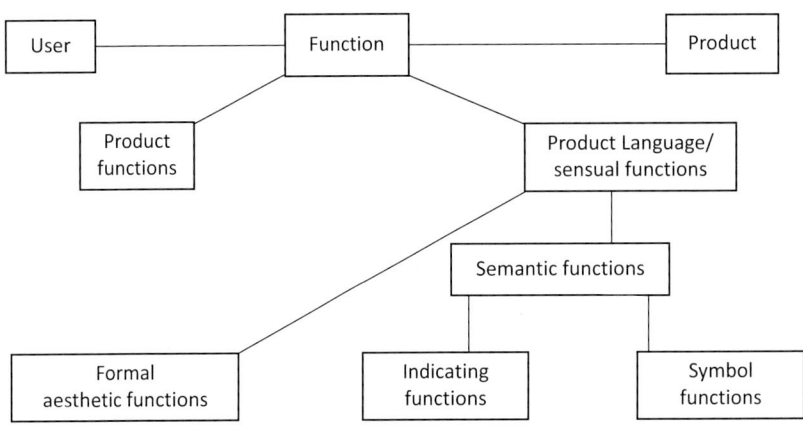

5.14
Conceptual model of the Offenbach theory of product language (modified from Gros, 1976).

178 □

like). There is inevitably a relation between product functions and the product language/sensual functions, so the two categories are not cleanly separable. Further, there is the possibility of a surplus of association or additional meanings that do not directly serve the user's needs. Among the designer's skills is to know how to tailor, tweak and massage forms so that there is the right amount of semantic functionality. That can only be done by a) paying close attention to the geometry, which means the curves, surfaces and proportions, and b) cross-checking how this is seen via user studies.

Nested under product language/sensual functions are two further aspects, 1) *formal aesthetic functions* and 2) *semantic functions*. While these are shown as two separate nodes, they are aspects of an object which exist in relation to each other. The 'form' in 'formal' directs our attention to the shape, colour and texture of the thing as we directly perceive it, or as if we can see the object without making conclusions of meaning (which essentially never, ever happens in real life).

Figure 5.15 expands on the category of formal aesthetic functions. It shows the range of options available to the designer in controlling the form, a choice between simplicity (left) or complexity (right).

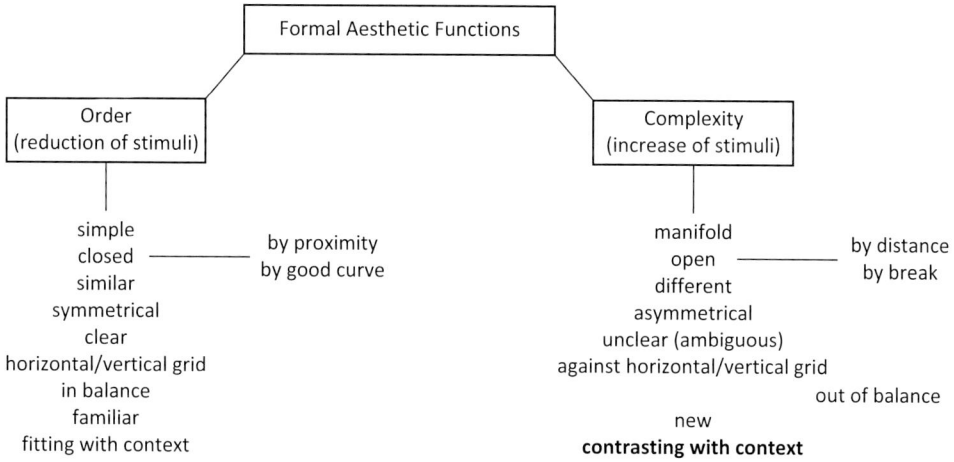

5.15
Conceptual model of the Offenbach theory of product language: the formal aesthetic function.

Under the rubric of semantic functions there is a distinction between 1) indication functions and 2) symbolic functions. The attempt to distinguish these is useful in that it makes one aware of the subtle differences in the purposes of forms and features. Indication functions (see Figure 5.16) seem very closely allied to affordance in that the user notices them in regard to understanding the basic purpose of the object. The directionality of a car or the way one façade looks more like the place to find a doorway than other ones: these would be the result of indication functions and are very much based on what the thing can do. There may be a close relation of the indication function and affordance, but it can be confounded intentionally and unintentionally.

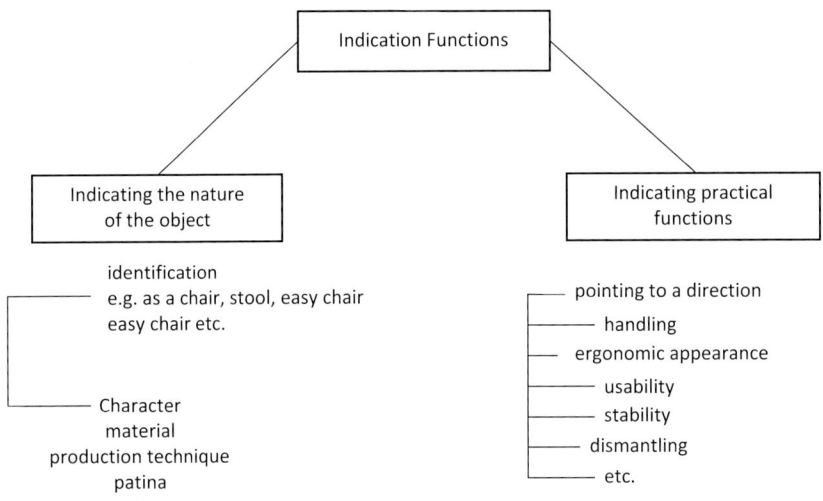

5.16
Conceptual model of the Offenbach theory of product language: the indication functions.

Symbol functions can be understood as geometries/colours/materials and textures so arranged that they convey a meaning. The three subcategories are shown in Figure 5.17: period style, partial style and associations. The associative category could be said to be related to the anthropomorphic/empirical or experience-based ways of looking. Although not included, masculine/feminine would be a natural candidate for the associations list. The category 'period style' indicates that Gros (1976) is suggesting that the style has a symbolic function as in when Victorian banks were built in classical style to suggest stability and strength. The category 'partial style' is a recognition of more vaguely defined but still identifiable genres of product styling. The items in 'partial style' could conceivably move into the 'period style' category with the passage of time.

A careful study of Figure 5.17 shows that it is not composed of three exclusive sets. For example, *under period styles* (left) is the subgroup 'functionalism' and in the middle under *partial styles* is 'German national style'. At times the two styles might be hard to distinguish. In some instances they may even be synonymous. The *period styles* subcategories could be more detailed. An architect might not wish to leave the subcategory 'classicism' unqualified if they were able to distinguish between German and, say, French classical. Further, 'classicism' could be associated with coldness or rationalism, items from the association group on the right. Dieter Rams' designs for Braun in the 1960s are modern and also now classical in the sense of being classical modernism.

Other cross-links between the groups in Figure 5.17 are possible. On the right side is the symbolic group 'associations' with the subgroup 'strong–weak', which is fourth in the list. Connoisseurs of English industrial design might see a characteristic of English design (a national style) as being its strength and robustness. That means that national style links to the design characteristic 'strong'. The same might go for German national style, which is also associated

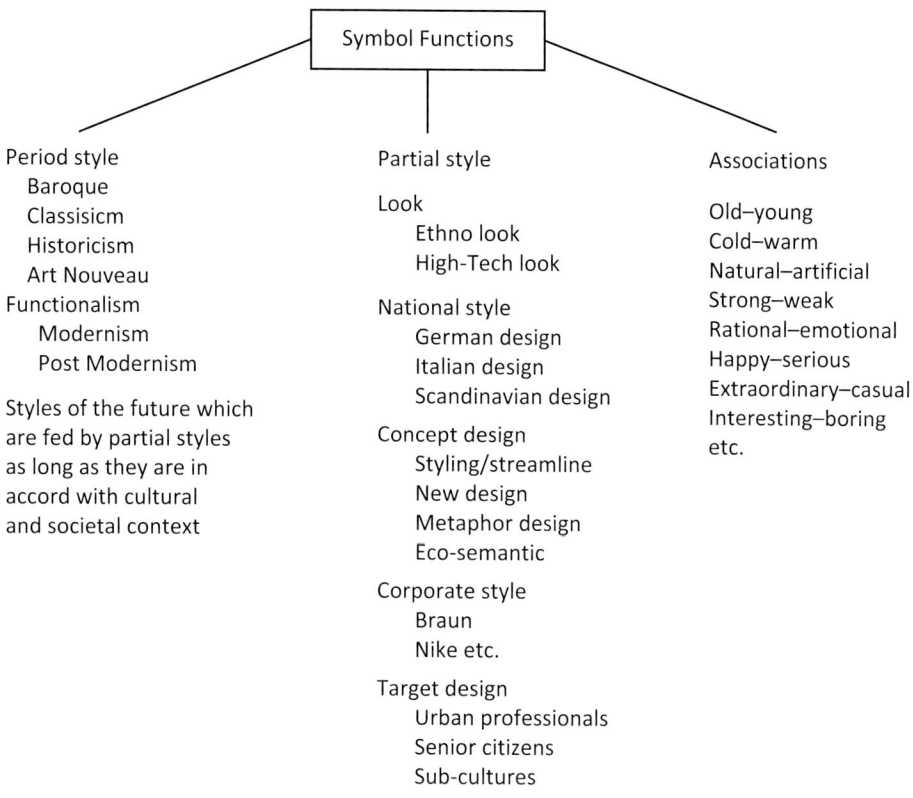

Symbol Functions

Period style
  Baroque
  Classisicm
  Historicism
  Art Nouveau
Functionalism
  Modernism
  Post Modernism

Styles of the future which
are fed by partial styles
as long as they are in
accord with cultural
and societal context

Partial style

Look
  Ethno look
  High-Tech look

National style
  German design
  Italian design
  Scandinavian design

Concept design
  Styling/streamline
  New design
  Metaphor design
  Eco-semantic

Corporate style
  Braun
  Nike etc.

Target design
  Urban professionals
  Senior citizens
  Sub-cultures

Associations

Old–young
Cold–warm
Natural–artificial
Strong–weak
Rational–emotional
Happy–serious
Extraordinary–casual
Interesting–boring
etc.

5.17
Conceptual model of the Offenbach theory of product language: the symbolic functions.

with robustness. So, in Figure 5.18, we can see a connection between the groups of ideas which are supposed to be, ideally, separate.

These questions are not answerable in a clear-cut way but point out the important sense that product semantics eventually rests on a good understanding of and sensitivity to a wide range of cultural influences and references. Unlike the analysis of curvature, semantics is about interpretation and there will always be a degree of subjectivity about this.

In this examination of the Offenbach model of product semantics we have thus seen that there are cross-links between the subcategories as in Figures 5.17 and 5.18. This reflects the many ways ideas can be put together and that associations are not hierarchical. From a product designer's point of view, the Offenbach theory of product language is not an excuse for ruminations on ontology.[3] It is a way to organise some broad categories of references so that certain aspects of the design can be analysed in (perhaps temporary) isolation.

We saw in Chapter 1 that there are several ways to account for aspects of human visual cognition. They were partial accounts and are also influenced by sociology as much as biology. In light of this, it should not be surprising that

■ **The meaning of the object and its elements**

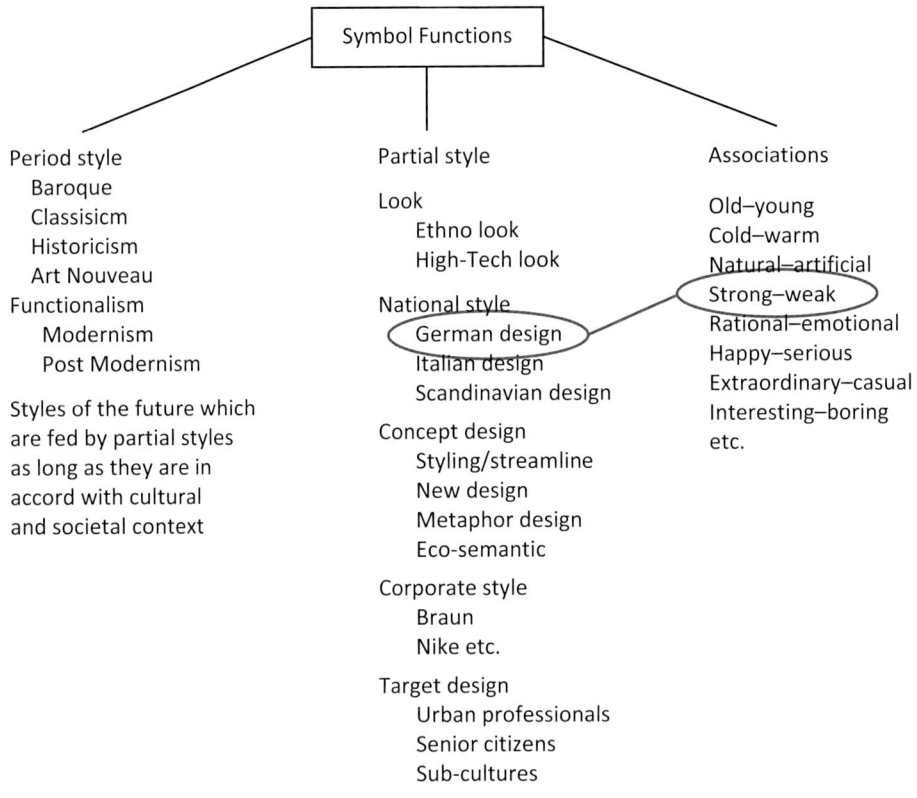

5.18
The connection between the partial style concepts and the associations concepts. Notice that the diagram shows concepts as being more clearly separable than they really are. That is not a failure, just a reflection of reality's complexity. Designed objects' meanings are multi-layered and the layers are not always distinct.

a semantic approach to objects' visual appearance should a) be partial like the accounts of visual cognition and b) not be amenable to very strict categorisation.

Finally, semantics are not just about rational analyses of objects. Semantic content can have emotional content such as nostalgia and excitement. Product semantics can also be used to help link to those things a person values.

EXERCISE 5.7

The aim of this exercise is to understand and apply the Offenbach model. The goal is to try and separate design elements in an object and fit them into the right place on the diagrams. It isn't really easy at first glance.

Apply the Offenbach model to a designed object. Here's an example to get you going. Consider a design for a suit. The conventions of a modern suit design relate to the reduction of stimuli. That means that a modern suit is a very simple garment in its appearance. The designer could add visual interest by having four different types of coloured buttons. That would be in contrast to the reduced, paired-down, sober forms of the classic suit. Referring to Figure 5.15 (left), picking coloured buttons is a formal aesthetic choice between order or variety. Furthermore, the effect of this choice relates to semantic functions (Figure 5.14) since a suit with odd buttons will be interpreted a certain way. The decision has meaning over and beyond the objective facts of the button choices. So, the choices under the category of formal aesthetic functions have a semantic impact as well as an aesthetic one.

Using this example as a model, think of another possible interlinkage of categories in the Offenbach model. To help you express the concepts, try to use words and images to make a graphic which displays the semantic breakdown of an object.

This may seem really theoretical. However, a solid grasp of this means you are able to manipulate associations and help the product to be understood in the right way.

5.6 USING PRODUCT SEMANTICS

Having outlined the main concepts in product semantics, the question remains of how to use this in the course of design projects. The underlying principle of semantics is that form and details of form are imbued with meaning. The scheme in Figure 5.19 shows how these concepts can be applied during research for design. At a basic level, the process deals with the search for the user's values with the aim of providing a product they can find meaning in. This

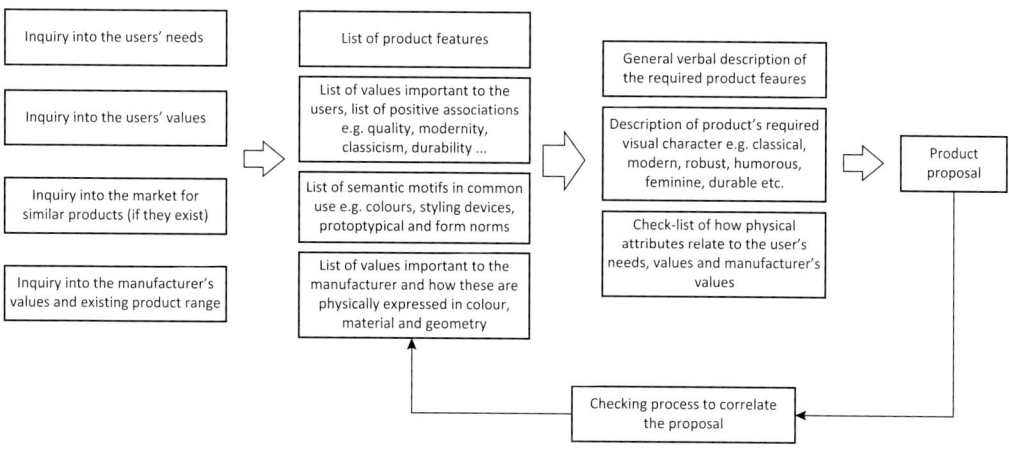

5.19
Gathering and using information relevant for a product's semantics.

is not to say that product semantics is about giving the consumer only what they know. Product innovation may involve the use of visual characteristics that are *not* in line with the user's known expectations. The product may have features and characteristics that do not conform to the norms of the product's class and subvert traditional, customary expectations. That might be an advantage in distinguishing the product in the market.

Figure 5.19 shows a general guide as to the main parameters related to gathering and managing information that a design team needs to map user values. The authors (Crilly et al., 2004, citing Butter, 1989) set out the task in the following way:

> The key stages of the process are: 1, establish the overall semantic character that the product should communicate; 2, list the desired attributes which should be expressed; and 3, search for tangible manifestations capable of projecting the desired attributes through the use of shape, material, texture and colour.
>
> (p. 562)

This is how that might be applied to hypothetical case of a portable smart speaker. The first group are semantic terms. The second group (2) is general attributes, and group 3 is a more specific interpretation.

1. Establish the overall semantic character that the product should communicate.

   Humour, gender-neutral, modern and technical

2. List the desired attributes which should be expressed.

   Use of distinct colours; anthropomorphism; the technical elements; organic form with some defined edges.

3. Search for tangible manifestations capable of projecting the desired attributes through the use of shape, material, texture and colour.

   Colour palette of warm grey with 021 C orange detailing; soft curves with gentle radii; orange fabric, soft-touch plastic; form should be asymmetrical but very stable, anthropomorphic buttons that look like eyes (for volume and on–off functions).

In this worked example, the designer might create two or three versions of list 2 and then proceed to makes sketches. Ideation is occurring in step 2, and tangible drawings and models come out in step 3. Then the results in step 3 can be compared to step 1 to see if they are in accord with the semantic goals.

This three-step process will help tie the functionality, geometry, colour and material of a product to the user's values. This opens a way to link individual lines, surfaces and geometrical forms back to concrete and definable values that the user is concerned with. Referring to Chapter 1 where the role of anthropomorphism is discussed, it is possible to put on one side the user's reported self-image and, on the other, the characteristics of the geometry that are animal- or human-like. The question for the designer is how the anthropomorphic aspects of the design accord with the user's self-image and product expectations. A very typical example is in the graphic treatment of a car's lamps and grille. These may be made to look more or less expressive. Currently there is a tendency for cars' front graphics to look aggressive. Customers presumably can relate to these forms. If they didn't, they would not be used. From the manufacturer's side, the product semantics (grille geometry, lamp geometry, etc.) are in also line with how the company wishes to present itself.

Finally, product semantics depends on good user investigations. That means that in order to use it effectively it must be carried out in tandem with activities like user interviews, focus groups and co-creation activities. This will generate a mass of data which has to be reliably linked to the tangible aspects of the product. That is a risky business and must be handled with care.

5.7 CRITICISM AND CONCERNS ABOUT PRODUCT SEMANTICS

Product semantics is not a precise business. There exist areas of ambiguity. Exactly how semantic elements are distinguished from ones that have a visual–cognitive basis is not clear. As has already been pointed out above, the degree to which semantic elements are separable from those related to affordance, gestalt theory and 'seeing as if' (see Chapter 1, Section 1.6) is open to discussion.

That the division of the elements of semantics are not fully defined is shown in the way the design researcher Nathan Crilly described the symbolic qualities of a product as *not* belonging to the category of social qualities. This is in relation to Crilly et al.'s (2004) discussion of how a product's appearance is used to represent the self to the user and to society around the user. Crilly et al. discounted symbolic qualities. But this chapter began with the examples of the two types of raincoat whose product semantics seems to be very much about communicating the user's identity to society, the symbolic function of the raincoats. This is another example of how there is potentially no definite semantic approach to design. That might lead some to question its validity.

The link between the features of the product and its meaning is not direct. At a basic and uninteresting level there is always the possibility that the customer is indifferent or they may misinterpret the selected proposal. Of more concern is mapping a clear channel from the way something is made to what it might mean. The gap between the symbolic and the concrete was described as follows by T.J. Van Rompay (2008, p. 333): 'although perceiving what products express comes most naturally, accounting for a product's expression is less straightforward'. In part this is because a large number of complex variables are involved in the relation – the user, the product and their social setting – and it changes over time. A fur coat today has a different meaning compared to 1965. Behind the product may stand a large set of expectations (positive and negative) as to what the product is like. Confirming the complexity of semantic interpretation, Van Rompay sought some explanations for the basis of the people's search for meaning. Two stand out for further discussion.

First is the matter of product expression and the empathy the user may feel for the item. The connection between the shape of the object and the feelings created by the viewer comes from the viewer interpreting the object. This is analogous to the way one interprets human body language. Upward body movements are associated with happiness and joy, while downward movements are associated with sadness. This explanation seems to be of a piece with the theory of anthropomorphism (see Chapter 1, Section 1.4); second, Van Rompay (2008) argued that affordances are involved in some of the meaning attached to forms. This would be obvious affordances like gripping, pushing, pulling which are readable on the object. I would argue that the affordances belong more at a lower level of basic product interpretation than semantics. So, basic things like being able to grip something are not semantic elements so much as basic physical attributes. It may be that affordances take on a derived semantic value; for example, the actual need for a low position in a racing bicycles evolves so that low seating signals sportiness whether sportiness is required or not. The role of affordance then depends on whether one judges the geometry to be firstly for functional purposes or whether the geometry is signalling a function or association that is not essential to the design. An example of the latter could be cartridge pockets on a jacket that is not intended for shooting.

Another ambiguity exists concerning the matter of geometry and style (where geometry refers to the shape of the object). Can you have geometry and not the style? Originally the modern movement involved an attempt to eliminate decorative elements in design for buildings and products. If we look at this example of Danish contemporary building (Figure 5.20), we see a building devoid of ornament and colour, in the modernist style.

This approach to architecture depends on using the bare minimum of elements. The individual items have no semantic content, having what Nygaard Folkmann (2013) would term zero surplus of meaning. Yet, as a whole, the building *has* semantic content: it attempts to convey notions of modernist austerity and cool rationalism. Paradoxically, the absence of decorative style lends the *entire* building semantic value. Meant as simply engineered apertures for light, the windows' appearance paradoxically become semantic tokens. Or are they? A case could be made that they are just glazed apertures minus any stylistic content. So, do these elements remain at the level of value-free geometry, or are they totems of a certain stylistic and design outlook? For some critics, even the fact the windows have 'landscape' rather than 'portrait' proportions would carry meaning; so, too, the absence of a pitched roof.

5.20
Contemporary building, Aarhus, Denmark, circa 2010.

Another case would be the simplified and cool forms of 1960s product design (Figure 5.21). Again, it was intended that these products had uncluttered and objectively rational shapes. They were designed supposedly to look very simple. The best design involves 'as little design as possible,' wrote the designer Dieter Rams (Vitsœ, 2021), which idea was clearly expressed on the 1963s Braun RT-20, which is from the same period. However, when the same approach is used by Apple for their products, it is hard not to see the forms

5.21
Grundig type 2029
radio, 1960.

as conforming to a style rather than being intuitive. Gros (see Section 5.5, Figure 5.17) placed style in the category of symbol functions under corporate style. The difference between cool, rationalist simple forms and the *style* of cool, rationalist simple forms does not lie in the object itself but in the mind of the user.

The author Susann Vihma (1995, p. 42) pointed out that Krippendorff and Butter's (1984) semantic concepts are so loose that they can be vague and too open in their meanings. As Vihma noted, Krippendorff and Butter (1984) even mentioned this: 'Because of this looseness, his approach to product semantics can end up being linked to marketing, which was not the purpose in the beginning' (Vihma, 1995, p. 42). This particular problem is not inherent in the concept of product semantics, though. All aspects of design can be linked to marketing, and the ultimate end of design is to lead to products which consumers wish to own and use. What Krippendorff and Butter (1984) are concerned with is designers cynically using product semantics to manipulate consumers. This problem, like all instances of bad design (as in unethical), can only by addressed by consumers who can judge if the symbolism of the product is aesthetically satisfying and consistent with the product's intended purpose. If they don't approve of it, they can elect not to buy the thing.

Researcher Ralf Weber (1995) discussed semantics in the context of design for buildings, but his critique can be applied to design generally. He addressed the concept of semantics under the general heading of theories of meaningful form, theories which claim:

> that the experience of any object rests on a mental image of it, a 'mental object' whose properties are not identical with those of the actual object. Form, according to this view cannot be naturally experienced; it always expresses or signifies something. In short, perceived form is always meaningful form.

Weber is sceptical about the ability of a designed object to 'represent, express, or even, like a language, communicate'. For Weber, meaning in designs is of three types: expressive, symbolic and semantic.[4] Weber disputed that objects can be expressive since they are inanimate. Representative art is expressive only inasmuch as it shows someone being expressive. For buildings (and presumably objects) the theory of empathy states that 'viewers project their own feelings into persons and objects and this endows them with expression'. Weber presented three reasons for why this might not be the case. Only the third is relevant, which is that often objects might not have any obvious interpretation at all.

Weber (1995) finally turned to theories of understanding design in semantic terms, by which he meant strictly linguistic parallels: 'Though there is no doubt that buildings [and designed objects] can evoke an indefinite and varying number of personal associations, these are not dependably specific semantic denotations that are required to make up a language'. In essence, words denote ideas by virtue of conventions, while things have meanings that 'are mere connotations'. In architecture and design the parts have no assigned, conventional meaning. And objects don't represent anything but themselves, argued Weber (1995):

> If architecture [and design] were a language, one could understand building [and objects] and control their meaning through design. For this to be the case, two conditions would have to be met: there would have to be clear semantic meanings for the constituents of architecture; and there would have to be a system of rules – a syntax – by which the meaning of the whole would derived from the meaning of the parts. Clearly, architecture meets neither condition.
>
> (p. 34)

The argument continues that most of the forms in design are meaningless, and if there are meanings, they are quite individual. The rules of some architectural styles are related to proportions. These rules of proportions are claimed to be analogous to syntax. In most or many cases the elements can be swapped about without affecting the proportions. In languages one can't swap the words and keep the grammar (i.e., the syntax). Weber (1995) concluded by saying meaning in design is something inferred by conventions, experience and learning. At best there is some intersubjective agreement that some shapes might mean something specific and no more than that. The core of Weber's argument, then, is that design's semantic content is uncertain. Talk of 'a design language' is just a metaphor or figure of speech and no more.

The philosopher Roger Scruton (1979) also argued for the limitations of the metaphor of language/signs on which the semantic approach to design rests. While Scruton wrote primarily about a semiotic approach to architecture, his point is valid for product design. His critique rests on the fact that a semiotic approach relies on knowing what language is and how it works in the first place:

> Architecture [*or design*] may resemble language either accidentally or essentially. It might share some or all of those features which go to make a language what it is, or it might share only those features which are linguistically dispensable; or again, it might share some of the essential features of language, but only by chance, as it were, and not through the very fact of being architecture. Only in the first case can we expect to derive from the linguistic analogy a theory of the understanding of architecture.
>
> (Scruton, 1979, p. 147)

It is, wrote Scruton (1979), only if architecture (or design) shares some of all of those features that make a language what it is that the linguistic analogy works. The two most important aspects of language that Scruton picked out are grammar and intention. Scruton conceded that designed objects (referring to architecture) display a kind of rule set or structure. Designers can talk of a consistent and well-resolved design, which means that the object displays a set of forms that conform to a finite set of rules for shaping the parts and their relation to each other. This is not the same as the grammar in language, which is 'necessary to show how the meaning of the parts determine the meaning of the whole' (Scruton, 1979, p. 152). A grammatically acceptable statement is 'connected to the possibility of truth' (Scruton, 1979, p. 152). In contrast, a grammatically unacceptable sentence in one language (e.g., 'cormorant only hat stand') has no truth content. A set of forms in a design, however well-composed, has no corresponding truth content either. The radio in Figure 5.21 has a style but is not making a statement of any kind. One can assemble design elements in a huge variety of combinations in a way that is unlike human language. Even if elements of a design can denote an idea – for example, Japanese tea-making or femininity – these are not truth statements. Perhaps at a very general level there might be the notion of 'honesty' in design such that a part looking like wood is made of real wood or a coat designed to look rugged is made of tough materials. This sense is quite weak (for philosophers) but probably serves designers well enough.

Scruton's (1979) and Weber's (1995) critiques serve to make one aware of pushing the semantic approach too far. What is left though is the idea of associations. Shapes have associations which lend them meaning. The totality of the associations that an object suggests adds up to a general cloud of meanings which, ideally, are consistent with each other even if the precise relation is not grammatically structured or amenable to tests of truth. While the proponents of the semantic approach try to push the concept of linguistic similarity too far, Scruton and Weber seem to strongly deny the value of any association which is, empirically, a bit of an overreach. Plainly, people *do* see associations in designed objects and derive meanings which will fundamentally affect the way they judge the object. Designers need to be aware of those associations and design with reference to them.

SUMMARY

This chapter examined how forms and elements of form can be seen as signs and to have a meaning beyond their pure geometrical content. Some researchers see strong parallels with language. In the highly elaborated system of the Offenbach theory it was apparent that the various levels of semantic meaning were cross-linked. A logically consistent hierarchy would not have this property. Nonetheless, designers can research users' values and create sets of associations that might be more attractive to the users. Critiques of the semantic approach drew attention to the ambiguity of semantic theory. They also talked about the weak line of reasoning leading from the basic stimulus (the geometry) to the meaning derived from it. Weber (1995) and Scruton (1979) challenged the claims that there was any linguistic parallel at all, given the lack of grammar (syntax) in designed objects.

This section concludes with the suggestions by Crilly et al. (2004) for further research:

> It would be beneficial to understand how designers incorporate visual references into their products. To what extent is this a conscious process and how are the visual references selected? Also, developing an understanding of the role of visual references in consumer response would be valuable. What is the range of references upon which consumers may draw and how might these best be categorised? To what extent are users aware of the references that are suggested by product form, and do they perceive the same references as those intended by designers?
>
> (p. 577)

Using some of the ideas in this chapter, each of these questions might be used to spur design work that results in forms that are more likely to be meaningful to the user.

NOTES

1 A point noted by Vihma (1995, p. 45).
2 The function, affordance and symbolism of elements of graphical user interfaces (GUIs) are not so easily divided.
3 Ontology is the study of categories. The tree of life diagram is an example of ontology, as is the way you organise the things in your fridge. It's worth looking into this topic since a lot of thinking depends on getting your categories right, which means ontological correctness!
4 The discussion of product semantics in this chapter has expression and symbolism as categories of semantics not distinct from product semantics.

FURTHER READING/KEY TEXTS

Crilly, N., Moultrie, J., & Clarkson, P.J. (2004) Seeing things: Consumer response to the visual domain in product design. *Design Studies*, 25(6), 547–577.

Scruton, R. (1979) *The Aesthetics of Architecture*. Princeton, NJ: Princeton University Press. **An essential book on the philosophy of architecture with much that is applicable to design**.

Vihma, S. (1995) *Products as Representations*. Helsinki: University of Art and Design.

References

Akner-Koler, C. (1994) *Three-Dimensional Visual Analysis*. Stockholm: University College for Arts, Crafts and Design.

Ashby, M.F., & Johnson, K. (2007) *Materials and Design: The Art and Science of Materials Selection in Product Design*. Oxford, UK: Butterworth Heinemann.

Behrens, R. (1998) Art, design and gestalt theory. *Leonardo*, 31(4), 299–303.

Bertamini, M., Palumbo, L., Gheorghes, T. N., & Galatsidas, M. (2015) Do observers like curvature or do they dislike angularity? *British Journal of Psychology*, 107(1), 154–178. doi: 10.111/bjop.12132.

Boess, S, & Kanis, H. (2008) Meaning in product use: a design perspective. In Schifferstein, H.N., & Hekkert, P. (Eds.). *Product Experience* (pp. 305–332). London: Elsevier.

Bridger, R.S. (1995) *An Introduction to Ergonomics*. Singapore: McGraw-Hill.

Broberg, O. (1997) Integrating ergonomics into the product development process. *International Journal of Industrial Ergonomics*, 19(4), 317–327.

Brown, D.C., & Blessing, L. (2005) The relationship between function and affordance. Paper presented at the ASME 2005 International Design Engineering Technical Conferences & Computers and Information Conference, Long Beach, CA, September 24–28.

Bruce, V., Green, P.R., & Gregory, M.A. (2003) *Visual Perception: Physiology, Psychology and Ecology*, 4th edition. Hove, UK: Psychology Press.

Confer, J.C., Easton, J.A., Fleischman, D.S., Goetz, C.D., Lewis, D.M.G., Perilloux, C., & Buss, D.M. (2010) Evolutionary psychology: Controversies, questions, prospects, and limitations. *American Psychologist*, 65(2), 110–126.

Cosmides, L., & Tooby, J. (1997) Evolutionary psychology: A primer. http://cogweb.ucla.edu/ep/EP-primer.html. Accessed May 18, 2018.

Crilly, N., Moultrie, J., & Clarkson, P.J. (2004) Seeing things: Consumer response to the visual domain in product design. *Design Studies*, 25(6), 547–577.

Dake, D. (2004) Aesthetics theory. In Smith, K.L., Moriarty, S., Kenney, K., & Barbatsis, G. (Eds.). *Handbook of Visual Communication: Theory, Methods, and Media* (p. 3). London: Routledge.

Dreyfus, H. (1967) *The Measure of Man: Human Factors in Design*. New York: The Whitney Library of Design.

Epley, N., Waytz, A., & Cacioppo, J.T. (2007) On seeing human: A three-factor theory of anthropomorphism. *Psychological Review*, 114, 864–886.

Føllesdal, D. (1997) Semantics and semiotics. In Chiara, M.L.D., Doets, K., Mundici, D., & Van Benthem, J. (Eds.). *Structures and Norms in Science*. Vol. 260 of *Synthese Library (Studies in Epistemology, Logic, Methodology, and Philosophy of Science)*. Dordrecht, The Netherlands: Springer. https://doi.org/10.1007/978-94-017-0538-7_28.

■ References

Frampton, C. (1995) *Studies in Tectonic Culture: The Poetics of Construction in Nineteenth- and Twentieth-Century Architecture.* Cambridge, MA: MIT Press.

Frayling, C. (2011) *On Craftsmanship.* London: Oberon.

Gibson, J.J. (1966) *The Senses Considered as Perceptual Systems.* Boston: Houghton Mifflin.

Gibson, J.J. (1972) A theory of direct visual perception. In Royce, J., & Rozeboom, W. (Eds.). *The Psychology of Knowing* (pp. 215–227). New York: Gordon & Breach.

Gibson, J.J. (1979) *The Ecological Approach to Visual Perception*, Classic edition. Hove, UK: Psychology Press.

Gomez-Puerto, G., Munar, E., & Nadal, M. (2016) Preference for curvature: A historical and conceptual framework. *Frontiers in Neuroscience*, 9, 712. doi: 10.3389/fnhum.2015.00712.

Gregory, R. (1970) *The Intelligent Eye.* London: Weidenfeld and Nicolson.

Grill-Spector, K., & Malach, R. (2004) The human visual cortex. *Annual Review of Neuroscience*, 27, 649–677.

Gros, J. (1976) Sinn-liche Funktion im Design. *Form, Zeitschrift für Gestaltung*, No. 74 and No. 75.

Gros, J. (1983) *Grundlagen einer Theorie der Produktsprache. Einfuhrung* [Fundamentals of a theory of Product Language. Introduction]. Heft 1. Offenbach, Germany: Hochschule für gestaltung Offenbach-am-Main.

Grossberg, S. (2014) How visual illusions illuminate complementary brain processes: Illusory depth from brightness and apparent motion of illusory contours. *Frontiers in Human Neuroscience*, 8, 854. doi: 10.3389/fnhum.2014.00854.

Han, S.H., & Hong, S.W. (2003) A systematic approach for coupling user satisfaction with product design. *Ergonomics*, 46(13–14), 1441–1461.

Haug, A. (2018) Acquiring materials knowledge in design education. *International Journal of Technical Design Education.* doi: 10.1007/s10798-018-9445-4.

Hegel, F., Eyssel, F.A., & Wrede, B. (2010) The social robot 'Flobi': Key concepts of industrial design. In *Proceedings of the 19th IEEE International Symposium in Robot and Human Interactive Communication (RO-MAN 2010)* (pp. 107–112).

Hekkert, P., & Karana, E. (2014) Designing material experience. In Karana, E., Pedgley, O., & Rognoli, V. (Eds.). *Materials Experience: Fundamentals of Materials and Design.* Amsterdam: Elsevier.

Jenkins, H.S. (2008) Gibson's 'affordances': Evolution of a pivotal concept. *Journal of Scientific Psychology*, 12, 34–45.

Kanis, H. (1998) Usage centred research for everyday product design. *Applied Ergonomics*, 29(1), 75–82.

Kant, I. (1928) *Critique of Judgment.* Oxford, UK: Meredith, Oxford University Press.

Karana, E., & Hekkert, P. (2010) User–material–product interrelationships in attributing meanings. *International Journal of Design*, 4(3), 43–51.

Kicherer, S. (1987) *Industriedesign als Leistungsberiech von Unternehmnen* (Industrial design as a service area of companies) (Diss.). GBI, Munich.

Krippendorff, K., & Butter, R. (1984) Product semantics: Exploring the symbolic qualities of form. *Innovation*, 3(2), 4–9.

Krippendorff, K., & Butter, R. (2008) Semantics: Meanings and contexts of artifacts. In Schifferstein, H.N., & Hekkert, P. (Eds.). *Product Experience* (pp. 353–376). London: Elsevier.

Lewis, W.G., & Narayan, C.V. (1993) Design and sizing of ergonomic handles for hand tools. *Applied Ergonomics*, 24(5), 351–356.

Liu, Y. (2003a) The aesthetic and the ethic dimensions of human factors and design. *Ergonomics*, 46(13–14), 1293–1305.

Liu, Y. (2003b) Engineering aesthetics and aesthetic ergonomics: Theoretical foundations and a dual-process research methodology. *Ergonomics*, 46(13–14), 1273–1292.

Loos, A. (1913) *Ornament et crime* [Ornament and crime]. Paris: Les Cahiers d'aujourd'hui.

Luna, B., Garver, K.E., Urban, T.A., Lazar, N.A., & Sweeney, J.A. (2004) Maturation of cognitive processes from late childhood to adulthood. *Child Development*, 75(5), 1357–1372.

Maier, J.R.A., & Fadel, G.M. (2007) Identifying affordances 2007. Paper presented at the International Conference on Engineering Design, ICED '0728, Cité des Sciences et de l'Industrie, Paris, August 31.

Michl, J. (1995) Form follows what? 1:50 - *Magazine of the Faculty of Architecture & Town Planning* [Technion, Israel Institute of Technology, Haifa], 10, 31–20.

Ming, C.C., Chien, C.C., & Chang., C. (2001) Perceptual factors underlying user preferences toward product form of mobile phones. *International Journal of Industrial Ergonomics*, 27(4), 247–258.

Monö, R. (1997) *Design for Product Understanding*. Stockholm: Liber.

Norman, D.A. (1988) *The Psychology of Everyday Things*. New York: Basic Books.

Nygaard Folkmann, M. (2013) *The Aesthetics of Imagination*. Cambridge, MA: MIT Press.

Ogden, C.K., & Richards, I.A. (1923) *The Meaning of Meaning: A Study of the Influence of Language upon Thought and of the Science of Symbolism* (Vol. 29). Harcourt, Brace.

Oliver, M. (2005) The problem with affordance. *E-Learning*, 2(4). doi: 10.2304/elea.2005.2.4.402.

Persson, S., & Wickman, C. (2004) Effects of industrial design and engineering design interplay: An empirical study on tolerance management in the automotive industry. Paper presented at the International Design Conference – Design 2004, Dubrovnik, Croatia, May 18–21.

Pinker, S. (1984) Visual cognition: An introduction. *Cognition*, 18(1–3), 1–63.

Post, R.A.G., Blijlevens, J., & Hekkert, P. (2013) The influence of unity-in-variety on aesthetic appreciation of car interiors. In *Consilience and Innovation in Design: Proceedings of the 5th International Congress of the International Association of Societies of Design Research* (pp. 1–6). Tokyo: Shabura Institute of Technology.

Pye, D. (1968) *The Nature and Art of Workmanship*. Cambridge, UK: Cambridge University Press.

Pye, D. (1978) *The Nature and Aesthetics of Design*. London: Herbert Press.

Scruton, R. (1979) *The Aesthetics of Architecture*. Princeton, NJ: Princeton University Press.

Sebestyen, G. (2003) *Construction: Craft to Industry*. London: Routledge.

Sowden, P.T., Davies, I.R.L., & Roling, P. (2000) Perceptual learning of the detection of features in X-ray images: A functional role for improvements in adults' visual sensitivity? *Journal of Experimental Psychology: Human Perception and Performance*, 26(1), 379–390.

Steffen, D. (2010) Design semantics of innovation: Product language as a reflection on technical innovation and socio-cultural change. In Vihma, S. (Ed.). *Design Semiotics in Use* (pp. 83–110). Helsinki: Aalto University.

Van Bezooyen, A. (2014) Materials driven design. In Karana, E., Pedgley, O., & Rognoli, V. (Eds.). *Materials Experience: Fundamentals of Materials and Design* (pp. 277–286). Oxford, UK: Butterworth-Heinemann.

Van Rompay, T.J. (2008) Product expression: Bridging the gap between the symbolic and the concrete. In Schifferstein, H.N., & Hekkert, P. (Eds.). *Product Experience* (pp. 333–351). London: Elsevier.

■ References

Vihma, S. (1995) *Products as Representations*. Helsinki: University of Art and Design.

Vitsœ (2021) The power of good design: Dieter Rams's ideology, engrained within Vitsœ. https://www.vitsoe.com/eu/about/good-design. Accessed November 3, 2021.

Ware, C. (2012) *Information Visualization: Perception for Design*, 3rd edition. London: Elsevier/Morgan & Kaufmann.

Waytz, A., Cacioppo, J., & Epley, N. (2010) Who sees human? The stability and importance of differences in anthropomorphism. *Perspectives on Psychological Science*, 5(3), 219–232.

Weber, R. (1995) *On the Aesthetics of Architecture: A Psychological Approach to the Structure and the Order of Perceived Architectural Space*. Aldershot, UK: Avebury.

Wertheimer, M. (1938a) Gestalt theory. In Ellis, W.D. (Ed.). *Source Book of Gestalt Psychology* (p. 2). London: Routledge & Kegan Paul.

Wertheimer, M. (1938b) Laws of organisation of perceptual forms. In Ellis, W.D. (Ed.). *A Source Book of Gestalt Psychology* (pp. 71–88). London: Routledge & Kegan Paul.

Westermann, J., Gardner, P., Sutherland, E., White, T., Jordan, K., Watts, D., & Wells, S. (2012) Product design: Preference for rounded versus angular design elements. *Psychology and Marketing*, 29(8), 595–605.

Wickman, C., & Söderberg, R. (2003) Increased concurrency between industrial and engineering design using CAT technology combined with virtual reality. *Concurrent Engineering*, 11(1), 7–15.

Index

acceleration (increase in curvature) 92, 93, 94, 99
aesthetics 7, 74, 97, 118, 120; and curve quality 78, 79, 92, 94; and geometry 88; and semantics 110
aesthetic values, compromisation of 51, 57
affordance 16–21, 27, 29, 47, 48, 59, 60, 168, 170, 172, 179, 185, 186, 191n2
affordance theory, critique of 19
Akner-Koler, C., three-dimensional analytical approach of 4, 39, 41, 42, 46, 47
alignment 151–153
anthropomorphism 4, 5, 30–34, 48, 184–186
Ashby, M.F. and Johnson, K. 63
axial space 88–90

Bauhaus 74; see also functionalism
Behrens, R. 8
Bertamini, M. and Palumbo, L. 79
Boess, S. and Kanis, H. 163
Bové, Lucian 106
brain functions 3
Braun 28, 28, 64, 180, 181, 187
Bridger, R. 59
Broberg, O. 59
Brown, D.C and Blessing, L. 17

cognitive system 3, 5, 7, 26; relation to theories of perception 17
colour break-up 71, 86, 87, 97, 118, 127, 129
Confer, J.C. et al. 32, 48
consistency 121, 156, 157–158; versus other values 161n3
constraints 50–75; as compromise 51–53; and functionality 53
context, effect on judgement 154–156

controlled convergence see parallelism
convexity, concavity 71, 113
corner, industrial design 76, 77, 110
corner, rounded see corner, industrial design
cost 50, 56, 69, 75, 75n1; as constraint 50, 51, 53, 57, 58; and craftsmanship 126, 137; and functionalism 71; and material 64, 68
craftsmanship and principal sections 121, 129, 131, 142; and semantics 153
Crilly, N. et al. 184, 185, 191
curvature 52, 63, 76, 78–91, 92–95; argument for quality of 78; and curvature continuity, compound forms 95–100, 103, 106; increase in see acceleration; mathematical description of 82–84; meaning of 78–80; and reflections 109, 110, 112, 113, 114, 115–116, 118; terms for 83
curve, transitional see fillets

Dake, D. 7
design values, design choices 51, 63, 72, 149, 161n4
Dibbern mug 81, 82
draft angle, drafting, draftability 63, 69 76n4, 126, 140, 146
Dreyfus, H. 59
Dunne, A. and Raby, F. 51, 75n2

ecological approach to visual perception 16
edge, industrial design 76, 77, 98, 111; see also fillets
electrolux Ultrasilencer 114
Epley, N. et al. 32, 48
ergonomics 6, 59, 60; and functionalism 73

evolutionary psychology 4, 5, 8, 17, 30–32

exploded view diagrams, value of 127, 128, 129

factor of closure 12, 13, 14; *see also* gestalt theory

factor of direction 12, 13; *see also* gestalt theory

factor of good curve 13, 15, 16, 179; *see also* gestalt theory

factor of proximity, 9, 10, 179; *see also* gestalt theory

factor of similarity 10, 11; *see also* gestalt theory

factor of uniform destiny (factor of common fate) 11; *see also* gestalt theory

feasibility 57

figure-ground 8, 9; *see also* gestalt theory

fillets 98, 103, 104, 119n2, 134, 142, 143, 146; *see also* edge, industrial design; rounded edges

Føllesdal, D. 166

forces (seeing 'as if') 4, 5, 39–47

form, compound 59, 78, 98–100; *see also* monovolume

form language 7, 54, 103, 111, 158, 169

Frampton, C. 123, 124, 125, 161n1

Frayling, C. 120

Fukasawa, N. 167, 173

functionalism 50, 51, 59, 71–74, 180, 181, 182; critique of 73–74; and mass production 73

functionality 6, 50, 51, 52, 53, 56, 57, 59–62, 76

functionality and functionalism 71

gender (perceived) 33, 34, 35; and anthropomorphism 4, 32; hazards of designing based on perceptions of 35

Georg Jensen 96, 97

gestalt, relation of Gibson and Gregory's theories to 17

gestalt theory 3, 4, 5, 8–15, 17, 23, 41, 42, 48, 90, 121, 147, 148, 174, 175, 185

Gibson, J.J. 4, 5, 8, 16–18, 19, 23, 27–29, 47–49, 59

glass 64, 65, 77

graphics 47, 117; applied 47, 76, 77, 117–118, 118, 170, 172; placement of 117

Gregory, R. 4, 5, 16, 17, 20–29, 48; critique of theory 24, 25

Grill-Spector, K. and Malach, R. 28

Gros, J. 178–183; *see also* product semantics, Offenbach theory

Grossberg, S. 3

Guidot, R. 65

Han, S.H. and Hong. S.W. 59

handcraft and mass production 50, 69–71, 73

Haug, A. 63

Hegel, F. 32

Hekkert, P. and Karana, E. 57, 63, 67

Hogarth, W. 79

homogeneity *see* consistency

injection moulding 54, 55, 71, 75n4, 120, 126

intersubjectivity 84, 159, 189

Jenkins, H.S. 19

joint lines 77, 86, 87, 114–116, 118, 127, 131, 134, 142; arguments for and against 129; classes of 134–137; construction of, problems with 114, 127; flushness of 137, 148; principle requirements of 126; and principle sections 129; quality of 114

Kanis, H. 59

Kant, I. 97, 118

Kettle 9, 29, 45, 47, 82, 173, 174

Kicherer, S. 174, 175

Koffka, K. 8

Köhler, W. 8

Krippendorff, K. and Butter, R. 163, 165, 166, 167, 168, 170, 172, 178, 188

Lewis, W.G. and Narayan. C.V. 59

line of beauty *see* curvature

line quality *see* curvature

lines 76–119, 125, 126, 131, 150, 160, 171, 174; panel gaps and shutlines as 137, 138, 139, 142, 143, 144–145, 147, 148

Liu, Y. 59

Loos, A. 72, 73

Luna, B. et al. 7

Maier, J.R.A and Fadel, G.M. 17

mass production and functionalism *see* functionalism and mass production material 50, 51, 52, 53, 56, 75, 97, 111, 161n4; and appearance 52; and constraints 52, 57, 63–69; and craftsmanship 140; and drafting 63, 69; effect on joints and junctions of 122, 123; semantics of 77, 126, 163, 164, 171, 173, 175–177, 178, 180, 183, 184

metal 64, 65, 66, 67, 69, 75, 77, 110, 132, 158; see also material, and drafting

Michl, J. 73, 75

Ming, C.C., Chien, C.C., and Chang., C. 81

Monö, R. 165, 172, 173

monovolume 99

Müller, Gerd A. (Braun designer) 28

Necker cube 23, 25, 26

Norman, D.A. 16, 18, 17

Nygaard Folkmann, M. 7, 27, 44, 47, 82, 186

Offenbach model of product semantics, analysis of 178–180

Ogden, C.K and Richards, I.A. 166

Oliver, M. 19

ontology 181, 191n3

order of assembly, effect of 148–150

ornament, as crime 72

panel gaps 137–148; radiused edges' effect on 143–148; "ratholes" due to 147

parallelism 150–151

Penrose triangle 26

perception, cognitive model of 3, 4; indirect and indirect see Gibson, J.J.; Gregory, R.; interference with 97, 124, 147; and light and shade 112; maturation of 7; semantic model of 176

plastic 39, 50, 54, 55, 56, 64, 65; effect on form of 63, 69; metal coating of 66, 67, 68, 69

Playmobil 120

Pollock, Ian 94, 95, 97

positive and negative spaces/forms 76, 77, 104–105, 110, 113

Post, R.A.G. et al. 157

Prägnanzstufen (law of simplicity) 11; see also gestalt theory

primitive forms (primitives) 4, 5, 35–39, 48

product semantics 34, 121, 162; communicated through (Boess and Kanis) 162; criticism of 164, 185–188 see also Weber, R.; Scruton, R.; definition of 163, 165; and gender 34; justification of 163–164; Krippendorff and Butter's theory 165–168; Offenbach theory of 178–183; structure of theory (semantic triangle concept) 166; using 163, 164, 183–185; Vihma's discussion of 173, 188

Pye, D. 51, 74, 119n1

quality 50–53, 54, 57, 58, 69, 70, 75n1, 130; perceived 126, 138, 148

quality and craftsmanship 120, 121, 122

quality and functionalism 71, 73

quality and material 63, 97, 132

Rams, D. 35, 180, 187

reflections (highlights) 76, 77, 78, 85, 88, 92, 93, 95, 99, 109, 113

robustness (firmitatis) 153–154, 154

rounded edges (industrial design edges) 54, 56, 63, 71, 98, 99, 101, 103, 111, 133; see also fillets

Scruton, R. 163, 189, 190

Sebestyan, G. 123

semantic design approach 162

semantic functions 178, 178, 179, 183

semantic perception, Kicherer's model of 175

semantic triangle 164, 166–168, 171, 174

semantics see product semantics

shadow, elongating effect of 108; see also positive and negative spaces/forms

shutlines see panel gaps

Sowden, P.T. et al. 7

space, positive and negative 76, 77, 104–110, 105

Steffen, D. 178

style see form language

styling, decoration 58, 61, 71, 72, 74, 103, 158

styling and semantics 180, 181

surface, transitional see fillets

surface quality see curvature

surfaces 76–119

tectonics 125

unity-in-variety 157
useless work (concept of) 74, 119 n1;
 see also Pye, D.

values of users 169, 173, 182, 183–184,
 183, 190
Van Bezooyen, A. 63
Van Rompay, T.J. 186
Vihma, S. 163, 163, 173, 188, 191n1
visual perception 3, 7, 16, 28, 164
Vitruvius 153, 160
Vitsœ 187

Ware, C. 19, 48 n1
Wassily chair 61
Waytz, A. et al. 30
Weber, R. 80, 163, 188, 189, 190
Wertheimer, M. 8
Westermann, J. et al. 81
Wickman, C. and Soderberg,
 R. 126
wood 56, 64, 65, 68; and
 production methods 69; semantics
 of 171, 190
wood-effect plastic, use of 66–67